GALILEO'S
SHADOW

The Theory Evolution in the
United States Courts

FREDERICK SPROULL

PAGE PUBLISHING
Conneaut Lake, PA

First originally published by Page Publishing 2023

ISBN 978-1-6624-8823-8 (pbk)
ISBN 978-1-6624-7710-2 (digital)

Printed in the United States of America

CONTENTS

PREFACE

I have had an interest in evolution since my early high school years. I was fortunate that in my high school biology class, the topic was addressed in an intelligent and sensitive way, unimpeded by politics and school administrators. As we will see in this work, that has not always been the situation in the United States. Although, evolution in my high school biology class made up only a very small part of the curriculum, the discussion caused my interest in the topic to increase and deepen. I must admit, however, that at that time, I had little understanding that evolution was so controversial, not scientifically, but politically. In college, my academic interest in evolution continued and increased as I took whatever advanced electives dealing with the topic that were available. Even at that early stage in my scientific journey, I began to understand that evolution was a special idea with incredible scientific, political, and religious implications—a theory that formed a framework for all of modern biology. Yet, it was only in law school that I first encountered, many years after it was decided, the 1968 *Epperson v. Arkansas* Supreme Court of the United States case addressing a creationist religious challenge to the teaching of evolution in public schools. *Epperson* marked the beginning of a line of federal court cases on the subject. Analysis of *Epperson* and the federal cases following it dealing with religiously based challenges to evolution in public schools, is the story told here. The interplay among the scientific, legal, and religious issues woven throughout the various court decisions are examined in this work and used as a model for better understanding the nature of science and how the judicial and legal systems often operate in the United States. Suggestions are made at the end of this work in an effort to accommodate both sides of this controversy.

The story about how Galileo Galilei's heliocentric views that the sun is the center of the solar system caused him to run afoul of the religious and political authorities of his day is well-known. Galileo was prosecuted by the Church's Court of Inquisition in 1633 and found guilty of "vehement suspicion of heresy." However, as a result of a plea bargain, he was not killed or tortured but sentenced to house arrest for life. Galileo was also made to renounce his sun-centered vision of the solar system, and he was forbidden from promulgating it going forward. Many interpret Galileo's ordeal as a classic confrontation between science and religion; a confrontation between the emerging modern science of astronomy and a particular religious interpretation about the universe. It is commonly seen as an example of religious bigotry and superstition attempting to suppress scientific progress and discovery. However, recent analysis of the facts reveal that this interpretation is only a part of a much larger and complicated tapestry: one full of duplicity and political, religious, and legal intrigue on the part of all concerned and governed by desires to gain or maintain political and religious power or scientific hegemony. These issues have been well documented in a number of places and it is not a goal to do so here.[1, 2, 3, 4]

A similar story, in modern trappings, is told here as science, religion, law, and politics are once again interwoven within a complicated political and legal milieu over three hundred years after Galileo. This story recounts a fascinating twentieth- and twenty-first-century creationist confrontation with evolutionary theory, not in the laboratory or field, but in the legislatures and courtrooms.

The story begins in the United States during the second decade of the twentieth century and continues into the twenty-first. In this story, evolution, not Galileo's astronomy, is the subject of judicial proceedings as Fundamentalist creationists challenged its legitimacy not just from a religious perspective, but also legally, politically, and even scientifically. The conflict centers on whether evolution should be included in the public-school science curriculum, and if so, how. This work tells this story through the lens of a number of salient court cases, mostly in the federal court system.

As with the Galileo matter, these evolution cases are also much more complicated and nuanced than merely a confrontation between science versus religion. They also represent a play for, or attempt to protect political and religious power or scientific priority. Discussion of these issues is woven throughout this work. The first of these cases is famous and part of American cultural lore, *Tennessee v. Scopes*.[5] The *Scopes* case was loosely, and in many ways fictionally, portrayed in the iconic Broadway play and movie *Inherit the Wind*. Although *Scopes* was tried in the state courts of Tennessee, it was followed by numerous federal court cases dealing with the teaching of evolution in public school science classes. Astonishingly, among these were two United States Supreme Court proceedings. However, before we go further, let's take a look at what the science of evolution is and what it attempts to explain.

Evolution is one of the most important concepts in modern biology. It's the scientific explanation for how organisms, including humans, got here. Every culture has a story about origins; evolution is biology and science's story. This has been true since Darwin published *On the Origin of Species*[6] in November 1859.

There are different ways evolution can be defined. The most basic and historical definition is that species can transmutate (change) over time into different species. It rejects a static view of the history of life that prevailed for most of Western history; it represents life as constantly changing through time. An evolutionary interpretation of the history of life sees all organisms as genetically related and linked through time.[7]

Evolution is essential to modern biology as its most bedrock unifying concept; its importance cannot be overstated. Reflecting on this, the great twentieth-century geneticist and evolutionist Theodosius Dobzhansky wrote, "Nothing in biology makes sense except in light of evolution."[8] Biology, without evolution, is not a cohesive science;

rather, it is only a catalog of unrelated facts that describe the living world. Evolution acts as an explanatory and organizing framework which allowed biology to emerge as modern science.[7]

Evolution also provides a relevant naturalistic, nonsupernatural explanation for the origin of different species. It frees biology from theology and religion. It does this by explaining two seemingly contradictory characteristics of life that have engaged naturalists and observers of nature for thousands of years: life's underlying unity and its obvious incredible diversity.

There are millions of different life-forms that inhabit this planet. Consider the vast array of different species of organisms that live in the various and diverse ecosystems of earth, from the almost limitless array of microbes to large animals and plants, and everything in between. Life on earth is diverse! Evolution explains this diversity by asserting that, over time, species branch and multiply into an almost endless array of new species. Another way of saying this is that species are not fixed entities; new species emerge through time with different characters from preexisting ones. Yet naturalists know that this diversity of life thinly veils an underlying unity that all living organisms possess. Living organisms share many fundamental characteristics with one another. This unity can be observed on at least three biologically levels: the level of the organism, the level of the cell, and the level of the molecule. For example, it has been observed for thousands of years that different species share characteristics with one another on the level of the organism; structural and functional similarities abound. Five digits, body segmentation, and bilateral symmetry are just a few similarities commonly found in various kinds of organisms. There are many others, as well, found in even diverse organisms.[7]

However, over the last couple of hundred years, even greater unity has also been discovered at the cellular level. Cells were first discovered toward the end of the seventeenth century with the invention of the microscope. Observations for almost two hundred years allowed naturalists in the nineteenth century to inductively reason that all living organisms are composed of structural similar cells. This conclusion was codified into the modern cell theory in the nineteenth century. The cell theory holds that all living organisms are composed of cells, and that the basic structural unit of all life is the cell, whether the life-form is bacteria, human, or anything else. There has never been a living organism discovered that is not composed of one or more cells.[7]

Since the middle of the twentieth century, technological advances have allowed an even deeper and more fundamental unity to be recognized by scientists at the molecular level. For example, we now know that all organisms discovered so far use DNA as the primary genetic material and that they all have a nearly identical genetic code made from DNA. In addition, living organisms use ribonucleic acid (RNA) in protein synthesis and share remarkably similar biochemical processes for making proteins. All living organisms also share many similar, or nearly identical, metabolic pathways and processes. In summary, there is close biochemical and molecular similarity among living organisms on earth. Organisms as diverse as bacteria and humans show an incredible unity at this level. This molecular unity has been exploited over the last few decades with new techniques and technologies. For example, human insulin genes and human growth hormone genes are routinely genetically engineered into bacteria cells where they are expressed by the host bacteria with human protein products that are harvested for medical use. Who would have believed, even a few decades ago, that the biochemical machinery of bacteria is so similar to that found in human cells that bacteria can be genetically engineered to express certain human genes and actually make functional human proteins? The three levels of unity discussed are explained through evolution. The unity is the result of all life being genetically related and derived from common ancestors over time.[7]

But what of the obvious diversity we see in living organisms on earth? That also is explained through evolution. Evolution is about change through time. Species and populations subjected to different environmental conditions and constraints diverge over time through Darwinian and other evolutionary mechanisms. The end result is change and the emergence of new subspecies and species of organisms. The diversity is explained as the result of branching and divergence from common ancestors over time.[7]

Finally, evolution explains another fundamental characteristic of life that has intrigued naturalists for ages—the adaptation of organisms to their particular environments. Naturalists and philosophers have noticed for thousands of years that living organisms generally fit well into the particular environments in which they live. This is called adaptation. The fin of the fish or the flipper of the whale for swimming, the wing of the bat or bird for flying, and the paw of the cat for running, climbing, and predation are all examples of adaptation. Over the centuries, adaptation was often explained in supernatural terms. One common idea from the seventeenth century through Darwin's time was that adaptation was a manifestation of God's wisdom and goodness. This is called natural theology. Darwin turned these idea 180 degrees by explaining adaptation in evolutionary terms. Adaptation, to Darwin, was the result of materialistic mechanisms driving evolutionary change of populations, rather than the result of supernatural forces.[7] Populations of organisms and species become adapted to their environments through evolutionary change over time.

Despite its importance to biology, evolution has always been a concept much larger than its biological meaning. It has far-reaching implications in medicine, politics, and philosophy. It explains so many seemingly diverse things. Why can so many different crops and animals be produced through domestication (plant and animal breeding)? Why do bacteria become resistant to antibiotics over time, and why do insects become resistant to pesticides? Why do new variants of flu and coronaviruses emerge? The answer to all of these questions is evolution. Populations evolve over time, populations change, and populations become different over time. These are all explained by evolution.

Prior to *On the Origin of Species*, most intellectuals did not have an evolutionary view of nature. However, within twelve years after its publication, evolution became a dominant intellectual paradigm in the West. This change in thinking has enormous significance and from its beginnings went far beyond biology. Freud believed that evolution represents one of two great intellectual revolutions of Western thought. It was only equaled, he believed, by the replacement of a geocentric model of the universe with a Copernican heliocentric one in the sixteenth and seventeen centuries. Freud felt that evolution was important and revolutionary because it removed humans from their status as specially created by God. Humans became merely one branch of the animal kingdom. Freud suggested that evolution's impact on how humans see themselves is qualitatively different than that of any other intellectual idea. It's this aspect of evolution that has generated so much religious and intellectual conflict over the last couple of hundred years and caused it to be part of numerous court proceedings in the United States during the twentieth century and twenty-first century.[7, 11]

In the last half of the twentieth century, Richard Lewontin also discussed evolution's broader importance. Lewontin stated that "[T]here have been only two real revolutions in biology since the Renaissance. The first was the introduction of mechanical biology by William Harvey and René Descartes. The second biological revolution, to which we attach the name of Darwin, is still being consolidated. Although its manifesto, On Origin of Species, appeared in 1859, it was not until the 1940s that Darwinism really established a hegemonic hold on such branches of biology as classification, physiology, anatomy, and genetics. It is still under siege by the restorationist armies of creationism, while at the same time it is undergoing a severe internal struggle to define its own orthodoxy and to resolve its own contradictions."[7, 9, 10]

The importance of evolutionary theory was also explored by biologist Ernst Mayr. Mayr lived a long life, dying in 2005 at the age of one hundred, intellectually active to the end. He was one of the twentieth century's outstanding evolutionary biologists and one of the intellectual parents of the new synthesis of evolutionary biology that occurred in the middle decades of the twentieth century. Mayr pointed out that evolutionary theory is so intellectually pervasive that almost every component of modern Western belief systems is "somehow affected by Darwinian principles."[12, 13, 14] His point was that ever since Darwin, evolutionary thinking has increasingly integrated itself into almost every part of Western intellectual life.

As might be expected from an idea that has such broad implications, evolution has often been used to support diverse and sometimes exploitative political and economic ideologies. In the nineteenth century and twentieth century, proponents of social Darwinism (a philosophy that attempted to [mis]apply evolutionary principles to human behavior, usually equating the fittest or best with those rich or politically powerful) and other exploitative philosophies such as laissez-faire capitalism, eugenics, Nazism, racism, and imperialism have used evolutionary theories, of one type or another, for support or ratification. In Darwin's time, biological evolution was used as a legitimating ideology supporting the bourgeois revolution. In this sense, it was used as scientific justification for the emerging mercantile capitalist class's attempt to gain political and economic power at the expense of the hereditary, static establishment.[7] Evolution is still being used to justify various ideologies. For example, beginning in the middle of the 1970s, proponents of a "new" discipline, sociobiology, proposes that Darwinian natural selection can explain a variety of complex social behaviors in humans including greed, selfishness, exploitation, racism, sexism, aggression, spite, xenophobia, conformity, and even upward mobility. The reason this idea is controversial is because complex human behaviors are seen as genetically determined. Human sociobiology has been challenged scientifically and ideologically as an extension of capitalist libertarianism and a modern form of social Darwinism, providing justification and validation for a variety of exploitative social behaviors.[7]

There has always been a bidirectional link between evolution and certain Western ideologies. Although evolution has served as an ideological foundation for various political philosophies, many argue that evolution itself emerged as part of a larger progressive ideology (progressivism) during the Enlightenment (seventeenth and eighteenth centuries), which explored in detail for the first time in Western thought the concept that change is natural and beneficial. Evolution was firmly linked, more or less, to progressive philosophy and ideology until the middle of the twentieth century when detachment from it began to become more mainstream. Evolution's detachment from progressivism has continued into the first part of the twenty-first century.[7] These ideas will be explored in more detail in this work.

No other scientific theory, at least in modern times, has provoked the intense religious and political opposition as evolution. Throughout history, no scientific theory has generated such intense religious opposition except for maybe Galileo's sun-centered cosmology. A main reason for this is because no scientific theory speaks so directly to human origins and who we are as evolution does. The religious implications of the theory are direct and palpable. As previously discussed, Freud understood the religious significance of evolution when he stated that evolution "dethrones humans from their lofty status as specially created beings."[10] More recently, Stephen J. Gould addressed this point by stating (overstating?) that "all thinking people accept the biological fact of evolution" and that "life on earth is not the result of special creation."[15] The tone of Gould's statement suggests that strong emotion exists on the scientific side of the fence on this issue as well as the side of certain religious groups when discussing evolution! Gould's view, if accepted, relegates over 40 percent of Americans into the class of "unthinking" according to recent polls![16, 17] Actually, most opposition to evolution since Darwin has not been scientific, but religious and political. Overwhelmingly, scientists have accepted some form of evolution. Few scientists, after publication of *On the Origin of Species,* rejected evolutionary explanations for life history. Notwithstanding this, everyone has heard of—or been part of—a religious group that formally opposes evolution. Opposition has been especially intense in Fundamentalist and Evangelical Christian religious communities in the United States. Why is this so? Especially since Darwin, the idea that organisms evolved and are evolving has generated minimal debate within the scientific community. Scientists have argued over mechanisms, mode, and tempo of evolution, sometimes with considerable rancor and much intensity but not whether it has occurred or is occurring.

So why has certain religious opposition been intense and unabated? This question is answered in the above paragraph.

Evolution, like no other scientific idea, treads on the heart of certain religious beliefs. This is because evolution does what no other scientific theory does or can do; it provides naturalistic answers to questions on how humans and other species got here. And it does this without having to rely on the supernatural. This puts evolution on a collision course with the religious ideas of many individuals and groups. This is what Freud meant.[7]

This is not to say that evolution is at odds with Western religions *per se*. Most mainstream Western religions, at least since Darwin, accommodate some form of it. Pope John Paul II provided an example of this accommodation when he stated that "new knowledge leads us to recognize that the theory of evolution is more than a hypothesis." However, he quickly limited his remarks and exposed the religious tension evolution engenders with the statement, "[i]f the human body has its origin in living material which preexists it, the spiritual soul is immediately created by God."[23] Actually, John Paul II reiterated Pope Pius XII's 1950 views in his encyclical *Humani Generis*.[18] As have Pope Benedict VI and Pope Francis. The view expressed by these various popes is more or less that of most mainstream, non-Fundamentalist-type religions in the West.[19, 20]

However, many modern biblical literalists (creationists), particularly in the United States, reject any accommodation, including ones that acknowledge the evolution of the human body, but not the human soul. These creationists often cast evolution as a naturalistic philosophy, irreconcilably at odds with religion, and certainly not science by any objective criteria.[7] The religious implications of biological evolutionary theory will be further explored in this work.

Since World War I antievolution creationists, mostly Christian Fundamentalists and Evangelicals, have become more politically important and influential, particularly in the United States. Many reject that evolution is even science; rather, they view it as naturalistic philosophy irreconcilably at odds with their religion and even the foundation of unsavory political and social philosophies.[7, 21, 22, 23] These creationists espouse a hyperliteral reading of *Genesis*.

Creationist opposition to evolution has especially focused and centered on the inclusion of evolution into the public-school science curriculum. In the 1920s, creationists came up with a novel approach to combat this. They began using the political and legal system to pass laws prohibiting the teaching of evolution.[24] This started a decades-long political and legal battle fought mainly, although not exclusively, in the federal courts, which in many ways continues to this day. No other scientific theory has ever been challenged or attacked in this way in the United States; nor has any other scientific theory been legislated against or been the subject of litigation.

Proponents of a form of creationism, known as scientific creationism or creation science, were particularly important and effective in these endeavors. Scientific creationism led the fight against evolution in the 1970s and 1980s, and in the guise of intelligent design it continued the fight into the twenty-first century. Scientific creationism emerged in the late 1960s, espousing two important antievolution beliefs. The first is a belief that earth's present geology is the result of a catastrophe (a worldwide flood, commonly known as Noah's flood). This contrasts with the view of most present-day evolutionists and other scientists who posit earth's geology as primarily the result of slow continuing naturalistic forces of change. Scientific creationists maintain that the fossil record reflects the flood and it is devoid of adequate transitional forms of life necessary to support evolutionary change.[7] These ideas are in sharp contrast to an evolutionary view of geology and biology held by almost all scientists since Darwin. To them, the fossil record represents a sequential appearance of evolutionary change of different species of plants and animals over a vast number of years of earth's history.[7]

A second antievolution tenet of scientific creationism is a belief that earth is young, only recently created by God (ten thousand years or less)[7], rather than the 4.6 billion years that most present-day geologists espouse.

However, for the first couple of decades of the twentieth century, most creationists or biblical literalists did not hold young-earth geological views. Many literalist Christians believed that the Bible allowed for an ancient earth and life before the Garden of Eden. They often accommodated modern geology by interpreting the six creation days of *Genesis* as representing vast ages in the history of the earth (day-age theory) or by separating an earlier initial creation from a later creation in six literal days (gap theory).[7]

Either idea allowed creationists to accept much of modern geology while defending the accuracy of the Bible. A very relevant example is William Jennings Bryan, the great United States populist politician, famous for his role in the *State of Tennessee v. Scopes* evolution trial, which will be considered next chapter. Bryan considered the creation days of *Genesis* to be geological ages.[7]

In addition to holding that the earth is only a few thousand years old and that the fossil record can be best explained by a biblical flood, scientific creationists raised a number of additional arguments against evolutionary theory. For example, they asserted that two great laws of physics, the first and second laws of thermodynamics, preclude evolution. The first law of thermodynamics states that the total energy in the universe is constant, neither increasing nor decreasing over time; the second law states that in the closed system of the universe energy tends to go from organized states to more disorganized states over time, ultimately ending in the most disordered form. Scientific creationists believe the laws of thermodynamics preclude evolution of highly organized living systems from less-organized ones. They believe that these great laws of physics do not allow complex organisms to evolve from simpler ones.[22] This interpretation of the laws of thermodynamics contradicts much of modern biology, chemistry, and physics. For example, modern physicists state that creationist interpretations of the second law apply only to closed systems, systems that do not gain energy from an outside source or loss energy to an external source. The earth is an open system with constant input of enormous quantities of energy mainly from the sun. In open systems there is no reason, according to the laws of thermodynamics, why solar energy cannot provide the energy necessary to drive complex chemical reactions necessary for life and evolution.[7]

Scientific creationists did not stop with a rewriting of modern physics. They also deny fundamental principles of biology. For example, the New Synthesis of evolutionary biology that occurred during the middle of the twentieth century, as modified by more recent discoveries, maintains that mutation, genetic recombination, and Darwinian natural selection are important in explaining much evolutionary change.[7] Scientific creationists dispute this, arguing that these processes cannot produce something as complex as an eye or a bird wing from simpler, less-complex structures by selection or other evolutionary mechanisms. Rather, they claim, in contradiction to almost all modern evolutionary thinking that complex structures cannot evolve from less-complex, simpler structures. Scientific creationists maintain that complex structures could not be favored by selection unless all the parts of the structure were in place at the same time, which they assert is impossible under an evolutionary view of life. In their view, it can only happen through creation of all the parts of the structure together.[7] Later in this work, scientific arguments refuting creationism will be discussed in detail.

Scientific creationists also seized upon the controversies among evolutionists surrounding evolutionary mechanisms to support their antievolutionary idea. They especially focused on the controversy surrounding the various forms of punctuated equilibrium, a theory of evolutionary change, and the assertions of certain philosophers contending that evolution is unscientific or not testable.[7] What is ironic in all this is that there is almost no disagreement among biologists that evolution occurred and is occurring; rather, any disagreements about evolution center on mechanisms, tempo, and mode of its occurrence.

Beginning in the late 1980s and into the twenty-first century a new form of scientific creationism, intelligent design (ID) emerged. ID is an idea attempting to explain the history of life. It specifically rejects modern evolutionary theory, asserting that the best explanation for the origin and diversity of life is the work of a designer, most likely of supernatural origin.

This work deals with the efforts of creationists during the twentieth and twenty-first centuries to eliminate or limit the teaching of evolution in public schools, and to include creationism in the scientific curriculum, the response these efforts generated from mainstream law, science, and religion in the United States. These efforts often resulted in political maneuvering, the enactment of laws, and litigation. Usually, the litigation found its way into the federal courts with United States Constitutional implications. No other scientific theory has ever been subject to such legal prohibition or judicial scrutiny in the history of this country. Astonishingly, twice, evolution cases even found their way to the Supreme Court of the United States. The interplay among the scientific, legal, and religious issues woven throughout the various court decisions that we examine in this work are of particular interest. Also of interest is how proponents of both evolution and creationism cynically altered and bent their arguments in response to previous court decisions in an effort to successfully promulgate their particular position and how courts often tailored their decisions to ideology, rather than judicial principles.

A story with scientific, political, and ethical implications is told in this work using a cluster of federal court cases that fairly represent the issues that are at stake in this controversy. Perhaps they also provide us with a model for better understanding how the judicial and legal systems operate in the United States. Suggestions are made at the end of this work in an effort to accommodate both sides of this controversy.

EVOLUTION STANDS FOR THE FIRST TIME BEFORE THE COURTS

State of Tennessee v. Scopes

In the United States there are two separate court systems that handle civil and criminal litigation. First, there is the judicial system of the federal government. Second, there is a separate judicial system sitting in each state. Although most evolution litigation has been in the federal system, we begin our discussion with the famous (infamous?) *State of Tennessee v. John Scopes*.[1] This was a Tennessee state case that marked the first time the legitimacy of evolution was considered by any United States court or, for that matter, the first time any scientific theory was challenged in this country in any court. Therefore, *Scopes* represents a unique event in the history of American science and jurisprudence. *Scopes* has become culturally iconic with numerous plays and movies over the years dramatizing true and fictionalized events from the case.

The story begins specifically in a state court sitting in Dayton, Tennessee. However, first, a bit of background is necessary before we begin. When Darwin published *On the Origin of Species*[2] in November 1859, he initiated an intellectual revolution in the West. Within a few years after publication of *Origin*, nearly every biologist of repute in America, Britain, Russia, Germany, and most of Western Europe accepted some form of evolutionary theory. Creationist explanations for life history were in severe decline among the intellectual mainstream.[3]

Evolution began to be incorporated into science textbooks in the United States in the last decades of the nineteenth century. At first, inclusion of evolution into textbooks did not generate controversy or legal issues. The first treatments of evolution in science textbooks for the most part attempted to reconcile the theory of evolution with divine creation. For example, Asa Gray, a prominent United States biologist in the 1857 edition of his book, *The Elements of Botany*, explained life history in terms of special creation by God. However, by the 1887 edition, Gray took a theistic evolutionary approach: that is, divinely directed evolution. In geology, we see a similar situation. James Dwight Dana's *Text-Book of Geology*, originally published in 1863, took a creationist viewpoint; however, the 1874 edition was revised to contain a limited version of theistic evolution. Further, although J. Dorman Steele's widely used science text, *Popular Zoology*, omitted evolution in the 1877 edition, he added a limited evolutionary view in an 1887 edition.[4]

A notable exception to this inclusive trend is found in *Principles of Zoology* by Harvard biologist Louis Agassiz. Agassiz was a leading United States zoologist who had emigrated from Europe. He was one of the very few United States scientists of repute to never accept some form of evolution. His antievolutionary views were strongly reflected in his textbook. From the 1848 edition to the final edition in 1873, the year of his death, Agassiz's *Principles* always presented his creationist, but nonbiblical, view that animal populations were specially created, where they subsequently lived, in a series of divine acts. These acts, Agassiz argued, occurred at intervals, over a very long earth history. Agassiz believed that massive extinctions punctuated this history into four distinct ages. He claimed that species were fixed, unchanging entities during each age, showing no evolutionary change. However, even Agassiz's genius could not turn back the evolutionary tide; all of his students accepted evolution. After 1880, United States science textbooks were almost always evolutionary in some form.[4]

This trend continued in the United States into the early twentieth century, but how evolution was addressed in science textbooks began to change. Theistic views were excluded. For example, Vernon Kellogg stated in his textbook, *Economic Zoology and Entomology*, in 1915, "Although there is much discussion of the causes of evolution there is practically none any longer of evolution itself. Organic evolution is a fact, demonstrated and accepted."[4]

By the early twentieth century, the textbooks even began to present human evolution as fact. Generally, these texts also completely omitted any reference to God. Several texts actually began to directly criticize creationist views. This represented a serious reversal from both the pre-*Origin* texts and the theistic evolutionary works of the latter part of the nineteenth century.[4]

Coinciding with this change was a tremendous expansion of the United States public-school system in the first decades of the twentieth century. As a result, nontheistic evolution was introduced to an increasing number of United States children, many of whom came from devoutly biblically literalist homes. And to exacerbate the situation, evolution was presented in a way that was directly critical of creationism.[3, 4, 5]

The groundwork was being laid for the political and legal opposition to evolution that began in post-World War I United States culminating in *Scopes*.[3, 4, 5] Opposition took a peculiar and novel form as antievolutionists turned to the political and legal system to eliminate evolution from the public-school curriculum. William Jennings Bryan, three-time presidential candidate and former secretary of state under President Wilson, was a prominent leader in the antievolution campaign. The campaign first manifested itself in the United States in the 1920s when antievolution legislation, barring the teaching of evolution in public schools, was introduced in the state legislatures of at least twenty different states; four states, Tennessee, Mississippi, Arkansas, and Oklahoma actually enacted such legislation into law.[4] In Tennessee, a 1925 law, the Butler Act, made it a crime to teach evolution in any public schools. That law, which is sometimes sarcastically referred to as the "monkey law" because of its emphatic opposition to human evolution, is the basis for *Tennessee v. Scopes*. *Scopes*[1, 4, 6] is the first of the many subsequent United States legal cases involving evolution in the public schools.

Now, let's return to the story of the *Scopes* trial. Soon after the enactment of the Tennessee's Butler Act in 1925, the American Civil Liberties Union (ACLU) placed an article in a Chattanooga,

Tennessee, newspaper announcing that it was seeking someone willing to test the act's constitutionality. The article attracted the attention of some civic leaders in nearby Dayton, Tennessee, who decided
that such a trial would bring attention and business to their community.[7] John Scopes, the athletic coach and physics teacher at the local
high school in Dayton, agreed to test the statute. Ironically, Scopes
was not even the regular biology teacher; he was substituting during
an end-of-term biology-class review. He did not even actively teach
evolution; his offense consisted of assigning pages, which discussed
evolutionary theory from a general biology textbook, *Hunter's Civic
Biology*). Scopes was arrested on May 7, 1925, and charged with
teaching the theory of evolution.[7, 8, 9]

The activities in Dayton attracted significant attention throughout the country. William Jennings Bryan volunteered to act as a special prosecutor for the State of Tennessee. Bryan had actively lobbied
for antievolution legislation. The ACLU accepted the services of
the famous trial lawyer Clarence Darrow to lead the legal defense of
John Scopes. Darrow was considered one of the outstanding criminal
defense attorneys of that time.[7, 8, 9, 10]

Darrow strongly objected to the Butler Act because he considered it the result of Fundamentalist religious fervor promoting
"ignorance and bigotry." Darrow thought that it was "bigotry for
public schools to teach only one theory of origins," referring to biblical creationism. Darrow's position on the Butler Act also mirrored
the position of the ACLU which stated, "attempts to maintain a
uniform orthodox opinion among teachers should be opposed" and
that "attempts of education authorities to inject into public schools
and colleges instruction propaganda in the interest of any particular theory of society to the exclusion of others should be opposed."
The ACLU sought to demonstrate that the Butler Act was unconstitutional because it made the Bible, a religious book, the standard
of truth in the public schools. Darrow also objected to the act on
constitutional grounds. He believed that it was an unconstitutional
infringement of religious freedom, due process, and free speech guar-
anteed by the Tennessee State Constitution.[7, 8, 9, 10, 11]

Bryan, on the other hand, supported the Butler Act as part of his populist politics. He believed that it represented a constitutionally permissible "determination of the parents to guard the religious welfare of their children" against the teachings of agnostics, atheists, and unbelievers. Bryan opposed teaching in public schools that the Bible was not literally true. He questioned whether "a minority in this state can come in and compel a teacher to teach that the Bible is not true and make parents of these children pay the expenses of the teacher to tell their children what these people believe is false and dangerous." However, despite his opposition to evolution, Bryan never supported the teaching of *Genesis* in public schools. He strongly believed that religion was not a morally or constitutionally proper subject for inclusion into the curriculum by legislation.[7, 8, 9, 10, 11]

Although Bryan was a religious Fundamentalist, his opposition to evolutionary theory was more nuanced than simply due to a belief in a literal interpretation of the Bible. Throughout his career he was a populist who deeply cared about equality and social and economic reform. He championed many progressive movements of his time. For example, he opposed American imperialism and the gold standard, and he supported such reforms as the popular election of senators, income tax, women's suffrage, and the rights of farmers and laborers.[11, 12]

Bryan actually had a rather permissive attitude toward evolution until World War I, when he became a vigorous opponent fighting hard for the passage of antievolution legislation. Bryan became involved in a campaign against evolution because he believed that evolutionary theory was being used by racists, militarists, nationalists, and supporters of certain supporters of abhorrent political and social movements. For example, the very book used by John Scopes, *Hunter's Civic Biology*, was blatantly racist, representing the Caucasian race as "finally, the highest type of all" and proposing involuntary sterilization as a remedy for crime and immorality. Throughout the 1920s, views of this type were not just confined to textbooks and scientific treatises. For example, in 1927, only two years after Bryan's death, and in the same year that the Tennessee Supreme Court decided the *Scopes* appeal, the eugenics movement in

America reached its zenith when the Supreme Court of the United States held in the infamous case, *Buck v. Bell*, that a 1924 Virginia law providing for the involuntary sexual sterilization of inmates of institutions was constitutionally permissible.[11,12, 13, 14, 15]

Carrie Buck, an eighteen-year-old woman, lived as an involuntary resident at the State Colony for Epileptics and Feeble-Minded in Virginia; she was committed there after birth of her illegitimate daughter. Carrie's mother preceded her as an involuntary resident. Carrie was the first person selected for sterilization under a new law. Justice Holmes delivered the infamous and chilling opinion of the court, allowing the forced sterilization to proceed stating, "Carrie Buck is a feeble-minded white woman who was committed to the State Colony... She is the daughter of a feeble-minded mother in the same institution, and the mother of an illegitimate feeble-minded child." Holmes continued, "three generations of imbeciles are enough."[13, 14, 15, 16]

Ironically, evidence gathered relatively recently has emerged showing that a key basis for the decision was not even true for a number of reasons. Most importantly, it has been ascertained that Carrie Buck was a woman of normal intelligence! Paul A. Lombardo of the School of Law at the University of Virginia, and a leading scholar of the *Buck v. Bell* case, wrote in a letter to evolutionist Stephen J. Gould, "As for Carrie, when I met her she was reading newspapers daily and joining a more literate friend to assist at regular bouts with the crossword puzzles. She was not a sophisticated woman, and lacked social graces, but mental health professionals who examined her in later life confirmed my impressions that she was neither mentally ill nor retarded.[13, 14, 15, 16]

Bryan was outraged by movements such as the eugenics movement. He believed that they were a direct attack on the morality and religion that had formed the basis of his entire political career. Bryan saw evolutionary theory being used as an ideological foundation and justification for them. In his last speech, published after his death, but originally written to be the closing address to the court during the *Scopes* trial, Bryan summarized his religious and political views on evolution. He contended that evolutionary theory destroys the religious beliefs of the people (especially children) in the literal interpretation of the Bible and in God, leading to corruption of morals, that evolution deadened the spiritual life of students to "love of God and love of fellowmen," and that evolutionary theory "would…carry man back to a struggle of tooth and claw." Further, Bryan contended that evolutionary theory paralyzed "the hope for reform" and discouraged "those who labor for improvement of man's condition." He linked evolution to the eugenics movement, which advocated sterilization of the "unfit" and "inferior." He associated evolutionary theory with political opposition to the building of asylums for the mentally ill and the sick and handicapped and with opposition to programs of vaccination. Bryan strongly supported these programs. He also linked evolution to the promotion of war and German militarism, which he believed contributed to the outbreak of World War I and with the promotion of social selfishness, "Each one for himself, and the devil take the hindmost."[11, 12]

Bryan's opposition to evolution was not simply the result of political ideology; it was also part of a larger Fundamentalist religious opposition. Following World War I, many Fundamentalists began to blame evolution for a perceived decline in morality. Their efforts to combat evolution, under the leadership of Bryan, led directly to the promotion of the laws prohibiting the teaching of evolution in the public schools, among which was the 1925 Butler Act.[8, 10, 11, 12]

Before World War I, however, by and large biblically literalist Christians and Fundamentalists were not particularly concerned with evolutionary theory as a threat to their faith. They certainly did not like evolution, but few saw the need to initiate a legal or political campaign to eliminate it from public-school curricula. Further, until rather recently (the last few decades), most creationists readily conceded that a literal interpretation of the Bible allowed for an ancient earth and perhaps life before the Garden of Eden. They generally accommodated the findings of historical geology either by interpreting the days of the first chapter of *Genesis* to represent vast ages in the history of the earth or by separating an initial *Genesis* creation "in the beginning" from a much later Edenic creation in six literal days. Either way, they widely believed that they could defend biblical accuracy while accommodating the latest geological and paleontological discoveries. Bryan, himself, interpreted the *Genesis* days as geological eras.[3]

The common view of modern creation scientists, who postulate that earth is very young (on the order of ten thousand years old), and who explain geological phenomena in terms of a Noah's flood gained significant support even in Fundamentalist circles only in the 1920s. The chief architect of this young-earth flood geology was George McCready Price who, during the early decades of the twentieth century, stood virtually alone insisting on the recent appearance of life and that the geological features of the earth were mostly dependent on a worldwide flood.[3]

Historian of science, Ronald Numbers, asserted that contemporary readers who associate creationism with the teaching of the modern scientific creationists will be surprised by the small number of nineteenth-century creationist writers who subscribed to a recent creation in six literal days, and the even greater rarity of those who attributed the paleontological findings in the fossil record to a worldwide flood. Creationists at that time, by today's standards, appear to be philosophically, theologically, and scientifically closer to theistic evolutionists than to the scientific creationists of today. It was not until the 1960s, with the publication of Whitcomb's and Morris' book, *Genesis Flood*, and the subsequent birth of the Creation Research Society, that Fundamentalists in large numbers began to interpret *Genesis* in a manner advocated by Price and to equate his views with orthodoxy.[3]

The development of Bryan's own attitude toward evolution closely paralleled that of the early twentieth-century Fundamentalists. As stated, before World War I, he was not overly concerned with evolution as a threat to Christianity or his values. He certainly did not see a need to launch a legal and political campaign to eliminate it from public-school curricula. Most significantly, he saw little reason to quarrel with those who disagreed with him on evolutionary issues. World War I, however, caused him to reevaluate his tolerant position. After the War, Bryan traced the source of many social ills to the influence of Darwinian thought.[3]

Bryan has often been criticized by historians of science for linking evolutionary theory, particularly Darwin's theory of natural selection, to certain political and social movements which Bryan found
abhorrent. Although this criticism of Bryan is reasonable in many
respects, the relationship of Darwinian evolution to unsavory early
twentieth-century political and social movements is nuanced and
complicated. For example, although the language used by industrial-
ists in Bryan's time to support *laissez-faire* capitalism appeared to be
Darwinian, it actually more closely reflected Victorian belief in the
virtues of self-help, and it was more an extension of liberal Protestant
morality than a strict application of Darwinian natural selection to
society.[5] Selection theory was co-opted to legitimate these ideologies
and political philosophies.

It has been further argued that justification of *laissez-faire* capitalism was actually more Lamarckian than Darwinian. Lamarckism
is an evolutionary theory that preceded Darwin's ideas. It was proposed by the French naturalist Lamarck in the very early nineteenth
century, which in pertinent part asserted that adaptive characteristics
can be acquired by exposure of organisms to a particular environment. Lamarck proposed that once acquired these characteristics can
be passed on to future generations. A classic example of Lamarckian
evolution is a herd of short-necked animals that find themselves in
an environment where the vegetation is mostly on high trees. The
animals stretch their necks to try to obtain the food. This constant
stretching causes their necks to elongate. They pass this change, or
acquired characteristic, on to their progeny, who also add to the elon-
gation in the same environment until long neck giraffes emerge.

This was very different mechanistically from Darwinian natural selection, which asserts that characteristics arise independent
of the environment in each generation and that they are preferen-
tially passed on to future generation only if they enhance survival
and reproduction in a struggle for existence among organisms in a
population. Characteristics that do not do so lead to the death of the
organism expressing them.

Let's use the same example to illustrate the difference between
a Darwinian view and a Lamarckian view of evolution. Once again,

we have our population of short-necked animals in living in an environment characterized by high vegetation on trees. Darwin recognized that within almost every population there is considerable variation. Therefore, our animals have various neck lengths. Animals with longer necks have more success retrieving vegetation and so, on average, live longer and produce more offspring. If neck length is genetic determined, they are likely to pass that trait on to the next generation. So as generations pass, the population of animals will have increasing longer necks on average due to differential survival and reproduction as the result of a favorable or adaptive trait, a longer neck. Darwin called this process natural selection, and it is his chief mechanism for evolutionary change. Darwin's animals are in a "struggle for existence" with one another with the result a statistically shorter life and fewer offspring for the shorter-necked variant. It must be stated that almost no scientists today hold Lamarckian views of evolution, but at the time of *Scopes* it was considered by many a major evolutionary mechanism. In fact, at the time of *Scopes*, more scientists held Lamarckian views than Darwinian ones.

Historian Peter Bowler contended that even the famous nineteenth-century "social Darwinist," Herbert Spencer, was at least as much a Lamarckian as he was Darwinian. Spencer believed that on a societal level, "struggle for existence" is not just a means of eliminating the congenitally unfit, but it is also a means of stimulating every individual in the population to maximum effort. The penalty for failure was not necessarily death; more often, the individual survived and learned from the experience how to do better next time. Struggle was the spur to self-improvement, and the accumulation of individual acts of self-development led to social progress. Therefore, the Darwinian concept of struggle for existence, within Spencer's philosophy, was linked to the Lamarckian mechanism of inheritance of acquired characteristics more strongly than to a Darwinian concept of selection and death. Spencer's "social Darwinism" was a really a form of "social Lamarckism instead."[17, 18]

It has also been argued that the eugenics movement was only indirectly dependent on Darwinian selection for its scientific or political foundation. Although it relied on the idea that relaxation of natural selection could lead to racial degeneration, it ultimately relied more on the application (misapplication!) of Mendelian genetic concepts for justification than on selection theory.[17, 18]

Nonetheless, despite these defenses of Darwinian natural selection's role in legitimizing abhorrent political and economic philosophies, it is undeniable that during the first decades of the twentieth century, evolution, including natural selection, was often used to justify a more ruthless approach to conquered and colonized peoples. Darwinian evolution was effectively co-opted to support the idea that various human races were strongly marked varieties of unequal ability. Bowler contended that Lamarckism had a similar history in Germany (Nazi race theory grew at least in part out of Haeckel's Lamarckian view of evolution) and in the United States. Evolutionary scientists of whatever variety in Bryan's day simply accepted—and many actively espoused—the pre-Darwinian racist view that non-Caucasian peoples were "inferior races," intermediate between apes and higher forms of humans (in their view, Caucasians).[17, 18] Although this idea predated Darwinism and reflected an ingrained Western cultural ideology, evolution of one type or another was integrated into the support structure legitimating these political ideologies.

It was within this scientific, social, and political context of post-World War I United States that the Tennessee Butler Act was enacted in 1925. The act prohibited "the teaching of evolution theory in all Universities, normal and other public schools of Tennessee, which are supported in whole or in part by the public school funds of the state, and to provide penalties for the violations thereof." The act provided that it was unlawful to teach in public schools "any theory that denies the story of the divine creation of man as taught in the Bible and to teach instead that man has descended from a lower order of animals.[19]

After a spirited trial in a Tennessee State court, John Scopes was convicted by a jury for violation of the Butler Act, and the trial judge imposed a fine of one hundred dollars. During the trial, the *Scopes* defense sought to introduce testimony from noted scientific experts addressing the legitimacy of evolutionary theory or addressing whether evolution contradicted the biblical account of creation.

The court sustained the prosecution's objection to such testimony and refused to allow it. The court stated that it was irrelevant to the legal issues involved. However, in a bizarre legal maneuver, Darrow called Bryan to testify as an expert witness on the Bible. It is of course legally unprecedented to call opposing counsel as a witness, especially for the other side to do so! However, the prosecution purposely failed to object, giving Bryan the opportunity he desired to challenge evolutionary theory. This set up a confrontation between Bryan and Darrow, two legal titans and two larger-than-life figures.[1, 7, 8, 9, 10, 11] Darrow's questioning of Bryan on the historical accuracy of the Bible and the legitimacy of evolutionary theory has become part of United States cultural mythology and urban history, inspiring numerous articles, plays, and movies. However, in reality, although the judge allowed Bryan to testify, the jury was not permitted by the court to hear it, and the testimony was later stricken from the official court record by the judge.[1, 7, 8, 9, 10, 11] Nonetheless, during the cross-examination, Bryan was humiliated by his contradictory answers, and he died few days after the trial ended.[20]

The judge was hardly unbiased. He was a conservative Christian, who began each day's court proceedings with prayer. However, the court justified its refusal to allow expert testimony on the basis of a politically expedient interpretation of the act. The court narrowly construed the Butler Act as only prohibiting the teaching that humans had descended from a lower order of animals,[1, 7, 8, 9, 10, 11], rather than interpreting it broadly as prohibiting any theory that denies the biblical account of the creation of man. If the court had chosen broad reading of the act, testimony on what constituted the biblical account and what constituted evolutionary theory (and whether evolution contradicted the biblical account) would certainly have been relevant and admissible. However, by narrowly limiting the boundaries of what the act prohibited, the court cleverly eliminated the Bible and evolution as relevant legal issues in the case. The narrow interpretation had the further effect of significantly undercutting any defense contention that the act was unconstitutionally vague in violation of constitutional due process requirements.[1, 7, 8, 9, 10, 11, 17, 18, 23]

The prosecution sought to limit expert scientific testimony in the case for good reason. Although Bryan rejected the scientific legitimacy of evolution, and although he had an early enthusiasm to turn the trial into a battle between religion and evolution, he was unsuccessful in procuring antievolution scientists to testify. Bryan was able to find certain conservative religious leaders willing to do so, but no scientists. Bryan contacted scientist George McCready Price, a noted creationist and proponent of flood geology, with a request to testify. Price acknowledged sympathy for the cause but refused, claiming he was in England at the time. Bryan found little support from others in the scientific community. The only respected scientist who agreed to come on Bryan's behalf was physician, Howard A. Kelly. Kelly warned, however, that although he rejected the concept of human evolution, he accepted the evolution of other species. Given this situation, the prosecution realized that it could not muster a scientific attack on evolution.[4, 11]

It may be added at this point that the conviction of John Scopes was appealed to the Tennessee Supreme Court by the defense.[1] In 1927, The Tennessee Supreme Court upheld the lower court's verdict, including its narrow interpretation of the act. The Supreme Court of Tennessee held, over defense objections, that the act was not unconstitutionally vague or uncertain, and as such it did not violate due process requirements of the Tennessee Constitution or the Due Process Clause of the Fourteenth Amendment to the federal constitution. The Tennessee Supreme Court further held that inasmuch as Scopes was an employee of the state of Tennessee, he had no right or privilege to serve the state "except upon such terms as the state prescribed. His liberty, his privilege, his immunity to teach and proclaim the theory of evolution, else than in the service of the state, was in no wise touched by this law." The Tennessee Supreme Court held in *Scopes* that the Butler Act was merely a valid exercise of a state's right to regulate the activity of employees while performing their job.[1]

The Tennessee Supreme Court also refused to consider the motives of the state legislature in passing the act. It held that the legal and natural effect of the act, rather than any legislative motive, determines the act's constitutionality.[1] However, as we will see throughout this work, the issue of legislative motive in determining an evolution statute's constitutionality is of primary importance. The *Scopes* court stands mostly alone with the position it takes on legislative motive. Subsequent to *Scopes*, motive of legislatures in enacting statutes is a key factor in federal courts' determination of the constitutionality of every evolution statute adjudicated.

Finally, the court held that the act did not violate the Tennessee State Constitution prohibiting any preference being given by law "to any religious establishment or mode of worship." The court justified this holding with an incredible statement, "So far as we know, there is no religious establishment or organized body that has in its creed or confession of faith any article denying or affirming such a theory... It would appear that members of the same churches quite generally disagree as to these things.[1] This is a standard that no subsequent court adjudicating a statute limiting the teaching of evolution in public schools ever used.

While upholding the constitutionality of the statute, the Tennessee Supreme Court overturned the lower-court conviction on a technical error inasmuch as the Tennessee Act required fines in excess of fifty dollars to be imposed by juries, rather than judges (a hundred-dollar fine was imposed on Scopes by the judge). In reversing the lower-court conviction, the Tennessee Supreme Court directed that since John Scopes was no longer employed by Tennessee, a "nolle prosequi" be entered in "this bizarre case" in the interests of "the peace and dignity of the state."[1] By overturning the conviction in this manner, the Tennessee Supreme Court dashed any hope or standing. The defense had to appeal the *Scopes* decision to the Supreme Court of the United States on constitutional grounds.

The *Scopes* trial occurred in the 1920s when the fact of evolution was no longer questioned to any appreciable extent in the professional scientific community. This is why Bryan had such trouble procuring expert scientific testimony. However, the actual mechanisms for evolutionary change remained intensely disputed at that time, as it still does in different ways today. Natural selection, Darwin's main mechanism for evolutionary change, had reached its lowest scientific status. With the exception of field biologists and biometrists, almost no scientists accepted natural selection as an important evolutionary mechanism. For example, paleontologists rejected natural selection in favor of Lamarckism or orthogenesis. Mendelians rejected natural selection in favor of mutation theory (also known as saltation theory).[1, 5, 17, 18]

Bryan attempted to use the scientific communities' uncertainty over the evolutionary mechanisms to his advantage. He questioned the value of a theory whose mechanism of action could not be elucidated with some degree of certainty.[4, 5, 10, 17, 18, 23] Bryan, of course, failed to anticipate the New Synthesis of evolutionary biology with its revival of selection theory in the middle of the twentieth century and the advances in molecular biology, genetics, and developmental biology in the last part of the twentieth century and twenty-first century, which would throw much light on evolutionary mechanisms. This is a story for the next chapter.

The legitimacy of evolution, as a scientific theory, was never directly put at issue in *Scopes*. Once the lower court interpreted the act narrowly, as only precluding the teaching that the human species evolved, it was able to uphold the act's constitutionality under the Tennessee State Constitution without such testimony. Looking back from the perspective of the twenty-first century, the *Scopes* court's justification rested upon the untenable constitutional conclusion that since members of various Christian churches differed in their views on evolution, the Tennessee State Constitution's prohibition on preference by the state on the establishment of religion or mode of worship was not violated. No significance was given by the *Scopes* court to the act's obvious preference and protection of a particular interpretation (Fundamentalist) of a Jewish/Christian holy book or, for that matter, to the preference given the holy book itself. The court also dismissed any obvious religious motive of the Tennessee legislature in passing the act as inconsequential to its constitutionality, a position that later federal courts would consistently reject.

The *Scopes* case never reached the federal court system or the United States Supreme Court as the defense had hoped. In a blatantly political maneuver, the Tennessee Supreme Court effectively put an end to the litigation by overruling the *Scopes* Lower Court on a technicality; therefore, the defense had no recourse to the federal court system.

Also, at the time of the *Scopes* decision, the religious clauses of the First Amendment to the United States Constitution, which we will see in subsequent chapters are so important to post-*Scopes* evolution litigation, were not applicable to the states and binding on state actions until after the 1940 United States Supreme Court decision in *Cantwell v. Connecticut*[21], and the 1947 United States Supreme Court decision in *Everson v. Board of Education*[22]. The First Amendment protections only applied to federal actions not those of the states until after these cases were decided. The religious clauses of the First Amendment of the United States Constitution are the Establishment Clause and the Free Exercise of Religion Clause. The First Amendment in pertinent part states that "Congress shall make no law respecting an establishment of religion or preventing the free exercise thereof." As a result, there was no specific legal precedent, at the time of the *Scopes* case, to support a federal claim that the Tennessee Statute violated the First Amendment to the federal constitution. There was, however, legal precedent at that time to argue that a vague state law violated the due process requirement of the Fourteenth Amendment to the federal constitution.[1, 7, 8, 9, 10, 11, 17, 18, 23] Darrow understood this distinction and argued this issue in *Scopes*, although unsuccessfully.

CHAPTER 2

EVOLUTION IN THE SUPREME COURT OF THE UNITED STATES

Epperson v. Arkansas

After the *Scopes* decision there was no litigation in the United States involving evolution for over forty years. Still, *Scopes* and Fundamentalist legal activity was not without consequences to the public-school curriculum. Publishers significantly deemphasized evolutionary theory in public-school curricula to avoid controversy. Generally, biology textbooks of that time period avoided discussion of evolution in contrast to the period immediately preceding the *Scopes* trial.[1,2]

In 1957, a seemingly unrelated event had profound implications for evolution in the public schools. The launch of the first artificial satellite, Sputnik, by the Soviet Union significantly changed the political climate. The United States government reacted strongly to the Soviet initiative with across-the-board efforts to modernize and improve the teaching of science in the public schools. The National Science Foundation began funding many different programs. Among them was an effort to develop new biology textbooks called the Biological Sciences Curriculum Study textbooks, or the BSCS texts. A major purpose of these books was to incorporate evolutionary theory as a major theme throughout the biology curriculum. The BSCS textbooks began to appear in the classrooms in the early 1960s.[1,2] These textbooks helped reawaken Fundamentalist fervor against evolution.

The wide incorporation of evolution into public-school biology textbooks in the 1960s contributed to renewed litigation some forty years after *Scopes*, culminating in the *Epperson v. Arkansas* case in the Supreme Court of the United State. The litigation began in 1965 in an Arkansas state court when Susan Epperson, a tenth-grade public-school biology teacher, brought legal action against the State of Arkansas. She demanded that an Arkansas antievolution law, which had been enacted in 1928, be declared void and that any dismissal attempt on her for violation of the statute be enjoined.[3] The Arkansas law was an adaptation of the Butler Act litigated in *Scopes*. There was a major difference, however. The Arkansas law deleted language found in the Butler Act, which prohibited teaching anything that "denies the story of the divine creation of man as taught in the Bible."[4, 5] This deletion was significant since the deleted clause generated important debate over statutory interpretation during the *Scopes* trial. The deletion of any reference to the Bible was an attempt to shield the Arkansas law from attacks that it favored a particular religion or type of worship, decreasing the chance that the law would be successfully challenged and ruled unconstitutional. Remember, the Butler Act was attacked in this way by the defense in *Scopes*; however, the *Scopes* court rejected this argument.

Here we see an example of a trend that characterized antievolution litigation throughout the years, each side of the controversy attempting to adapt its position to avoid the best arguments of the other side in a type of evolutionist/creationist arms race. Oftentimes, each side did this to such an extent that it violated the integrity of its own position. Even the courts were not exempt from behavior of this type as we will see.

The Arkansas law litigated in *Epperson* made it unlawful for a teacher in any state-supported school or university "to teach the theory or doctrine that mankind descended from a lower order of animals" or "for any teacher, textbook commission, or other authority exercising the power to select textbooks" for adoption in any public school to adopt or use a textbook that "teaches the theory or doctrine that mankind descended from a lower order of animals." Violation of the law was a misdemeanor and subjected the violator to dismissal.[4] As we have discussed, the language of the law was a thinly veiled attempt by Arkansas to avoid the religious problems raised by the

Butler Act, yet still eliminate evolution from the public-school science curriculum.

The events in this case were triggered when the official textbook for the high-school biology course in a Little Rock public-school system was changed for the academic year of 1965–1966 to a BSCS book containing a section on evolution. The prior textbook did not contain such a unit. Susan Epperson, at the beginning of the 1965–1966 academic year, was presented by state officials with the new book for use in her classroom. Technically, to do so would have been a criminal offense under the state's antievolution law. Nonetheless, Epperson's concerns were disingenuous to say the least. Even though the law was enacted in 1928, there was not even one attempt by Arkansas to enforce it, nor was there any indication that it would be enforced on Susan Epperson. In any event, Epperson instituted an action in a state Chancery Court. The Chancery Court in Arkansas held in an unreported opinion that the statute violated the First Amendment to the federal constitution.[3] This was a significantly different holding than the one in Tennessee by the *Scopes* court.

On appeal, the Chancery Court decision was reversed by the Arkansas State Supreme Court that held, like *Scopes*, that the law was a "valid exercise of the state's power to specify curriculum in its public schools."[7] The case was appealed to the Supreme Court of the United States where it was decided in 1968.[3] The *Epperson* case presented a situation where, for first time in United States legal history, evolutionary theory stood for judicial review in a federal court, let alone the highest court in the United States.

In *Epperson v. Arkansas*, the Supreme Court of the United States unanimously held that the Arkansas law was an unconstitutional violation of the First Amendment to the federal constitution. The court's rationale was that the law selected "from the body of knowledge a particular segment which it proscribes for the sole reason that it is deemed to conflict with a particular religious doctrine; that is, with a particular interpretation of the book of *Genesis* by a particular religious group."[3]

The *Epperson* decision directly contradicted the Tennessee Supreme Court's holding in *Scopes*. Remember, *Scopes* held that the

Butler Act did not give any preference "to any religious establishment or mode of worship.[7] As we will see, the *Epperson* view on this matter is the prevailing view various Federal courts took in subsequent similar litigation.

The Supreme Court arrived at its decision in *Epperson* by relying on important developments in First Amendment constitutional law that occurred subsequent to the *Scopes* case. The First Amendment, the first of ten amendments known as the Bill of Rights, to the federal constitution includes two, so-called religious clauses: the Establishment Clause and the Free Exercise Clause. These clauses, in pertinent part, mandate that "Congress shall make no law respecting an establishment of religion or preventing the free exercise thereof." However, when the Bill of Rights was ratified by the original thirteen states in the eighteenth century, it applied only to the federal government; it protected individuals only from the federal government actions. The Bill of Rights did not apply to state actions. This situation continued into the 1940s, and it was constitutional law at the time *Scopes* was decided.

The legal landscape began to change and culminated in the 1940s when the Supreme Court ruled that the Fourteenth Amendment to United States Constitution should be interpreted as applying the two First Amendment religious clauses to the states. This occurred in *Cantwell v. Connecticut* in 1940 and *Everson v. Board of Education* in 1947. Together, these cases explicitly made both religious clauses fully applicable to the states as well as the federal government. [3, 8] Relying on *Everson* reasoning, the *Epperson* court stated, "Government in our democracy, state and national, must be neutral in matters of religious theory, doctrine, and practice. It may not be hostile to any religion or to the advocacy of no-religion; and it may not aid, foster, or promote one religious theory against another or even against the militant opposite. The First Amendment does absolutely mandate government neutrality between religion and religion, and between religion and nonreligion." Two important and relevant ideas are contained in this quote. The First Amendment to the federal constitution applies to the individual states, not just the federal government, and the First Amendment requires the states to act neutrally toward religion.

The prevailing test for religious neutrality at the time *Epperson* was decided had been formulated in 1963 in the famous Supreme Court case, *Abington School District v. Schempp*[9], where devotional bible reading in public schools was constitutionally prohibited. The *Schempp* test required that neither the primary purpose nor the primary effect of a state action must advance or inhibit religion. Therefore, when the *Epperson* court reviewed the Arkansas antievolution law, its sole question was a very different one than what was before *Scopes*. It was whether the act's legislative purpose or effect promoted religion[3]! Remember, the *Scopes* court believed that purpose was irrelevant to its deliberation.

In applying the *Schempp* test to the Arkansas law, the *Epperson* court concluded that the law was enacted to promote the religious views of some of its citizens, violating religious neutrality. The court believed that Fundamentalist religious belief was the sole motivating purpose behind the Law. *Epperson* stated that, although Arkansas eliminated certain references to the "Divine Creation of man as taught in the Bible" from the Butler Act litigated in *Scopes*, the Arkansas law and its purpose was essentially the same as the Butler Act: that is, to suppress the teaching of a theory that denied the divine creation of humans. In reviewing the history of the law, the court also recounted that the Arkansas law was adopted by popular initiative in 1928, only three years after the Butler Act was enacted and only one year after the Tennessee Supreme Court's decision in the *Scopes* case.

The *Epperson* court also felt that the advertisements used in the campaign to secure adoption of the Law provided evidence for the Law's religious purpose. For example, an advertisement, cited by the court, read, "THE BIBLE OR ATHEISM WHICH?" All atheists favor evolution. If you agree with atheism vote against Act No. 1. If you agree with the Bible vote for Act No. 1... Shall conscientious church members be forced to pay taxes to support teachers to teach evolution which undermine the faith of their children? The Gazette said Russian Bolshevists laughed at Tennessee. True, and that sort will laugh at Arkansas. Who cares? Vote for Act No. 1."[3]

Based on its interpretation of the history and legislative history of the act[3], the *Epperson* court concluded that, although individual states and local authorities have broad discretion in how they operate public school systems, the courts must intervene to protect constitutional freedoms. Schools cannot violate constitutional rights and guarantees.

Therefore, the *Epperson* court reached a historic decision in First Amendment jurisprudence. For the first time in United States legal history, a prevailing constitutional test was used to invalidate a statute on First Amendment grounds because of the supposed motivation or purpose of a state legislature in enacting it.[3, 9, 10] After *Epperson*, nearly all antievolution laws litigated in the Federal Courts through the years were analyzed using a standard similar to the one articulated in *Epperson*.

The Supreme Court of the United States in *Epperson* and the Supreme Court of Tennessee in *Scopes* interpreted basically the same laws in vastly different ways. In *Scopes*, the Tennessee Supreme Court held that the Butler Act was merely a valid exercise of a state's right to regulate the activity of employees while performing their job. It dismissed claims that the act violated any prohibition regarding the establishment of religion or mode of worship, either by state or federal constitutions. In *Epperson*, however, the Supreme Court of the United States held that the Arkansas law violated the First Amendment of the United States Constitution. As such, any legitimate right a state might have to regulate employee activity must be subordinated.

As discussed above, it wasn't until after the 1940 United States Supreme Court decision in *Cantwell v. Connecticut* and the 1947 United States Supreme Court decision in *Everson v. Board of Education* that religious clauses of the First Amendment to the federal constitution were applied to the individual states.[3, 9, 10] As a result, the defense team during the *Scopes* trial did not have legal precedent necessary to effectively argue that the Butler Act violated the religious clauses of the First Amendment to the United States Constitution (an issue essential in *Epperson*). By the time *Epperson* was decided, however, it was a well-established constitutional law that the First Amendment applied to the states as well as the federal government. In addition, by 1963 through *Schempp*, a constitutional test for religious neutrality was formulated by the Supreme Court of the United States and that test was used by the *Epperson* court in determining that the Arkansas law did not meet First Amendment constitutional criteria. Under the requirements of this test, the primary purpose or the primary effect of a state action must neither advance nor inhibit religion. The *Scopes* court had specifically rejected any purpose analysis stating that only the effect of a statute, rather than any "proclaimed motives" of the legislature or the statute itself, determine an act's constitutionality. Much had changed legally in the intervening years between *Scopes* and *Epperson* decisions.

Although the *Epperson* decision was unanimous, not all members of the *Epperson* Supreme Court agreed with the reasoning behind the decision. Three justices, including Justice Hugo Black, disagreed with much of the majority reasoning and wrote separate, but concurring, opinions. All agreed that the Arkansas law should be stricken for vagueness, if for no other reason[3,11]

However, Black's opinion raised certain issues that are still argued. He rejected the court reviewing legislative motives behind the law. Black believed that the Arkansas law might infringe on the religious freedom of those who believe that evolution is an antireligious doctrine. However, he also believed that since Arkansas made no attempt to include a literal account of *Genesis* in the public-school curriculum, removing evolution might represent true religious neutrality under the First Amendment. Black stated that "Unless this Court is prepared simply to write off as pure nonsense the views of those who consider evolution an anti-religious doctrine, then this issue presents problems under the Establishment Clause far more troublesome than are discussed in the court's opinion." [3,11] Yet that is precisely what the *Epperson* court did. And as will be discussed in subsequent chapters, most federal courts since *Epperson* have done the same. Nonetheless, similar concerns were raised again over forty years later by Judge Scalia, writing in the late 1980s on behalf of the dissent, in the second Supreme Court case to address evolution in public schools, *Edwards v. Aguillard.*[23]

I don't believe that the changes in the legal landscape between *Scopes* and *Epperson* were the only reasons for the remarkably different decisions in the cases. Surely, the *Epperson* court was also influenced by conceptual and scientific advances that occurred in the field of evolutionary biology during the intervening forty years between the decisions.[12,13] As will be seen in subsequent chapters, almost all federal courts seem to have been greatly influenced by, and they seem to have accepted *carte blanch*, mainstream scientific opinion on evolution. It is one of the hallmarks of this line of federal cases.

Evolutionary biology as a scientific theory matured greatly between the time *Scopes* and *Epperson* were decided. During the first part of the twentieth century, the field of evolutionary biology was fragmented. Although evolution was viewed as "fact" by most biologists, there was enormous diversity of opinion as to what mechanisms are central to evolutionary change. There was little or no consensus in the scientific community on this matter.

This fragmentation was documented in William Bateson's famous address before the 1921 convention of the American Association for the Advancement of Science. He stated, "When students of other sciences ask us what is now currently believed about the origin of species, we have no clear answer to give." However, Bateson continued, "We have absolute certainty that new forms of life, new orders, new species have arisen on earth…our faith in evolution remains unshaken."[14] In other words, the idea that evolution is the best explanation for life history was widely accepted at the time of the *Scopes* decision, but there was no agreement on which mechanisms are important to evolutionary change.

For example, evolutionary biology in the early twentieth century consisted of many subdivisions, each with its own mechanism for explaining evolutionary change with little or no agreement with one another. Although Darwin convinced the intellectual world that evolution is the best explanation for life history with publication of *On the Origin of Species* in 1859, his main mechanism for evolutionary change, natural selection, did not do well. Soon after publication of *Origin* through almost the first half of the twentieth century, natural selection, with the exception of field biologists and biometrists, was rejected by evolutionary biologists as an important mechanism driving evolutionary change. Paleontologists rejected natural selection in favor of Lamarckism evolution or orthogenesis. Mendelians (experimental geneticists) rejected natural selection in favor of mutation theory (also known as saltation theory).[12, 13, 15, 16] Although mechanisms were in dispute, as already stated, evolution as a reality was accepted by almost all intellectuals. This was the state of evolution biology at the time of *Scopes*.

It was not until the 1930s, 1940s, and latter that the differences among the various subdivisions were narrowed to the point where a general consensus emerged among evolutionary scientists. This consensus is commonly termed the New Synthesis of Evolutionary Biology, the Evolutionary Synthesis, or the Modern Synthesis.[12, 13, 15, 16, 17] Although evolutionary biology has continued to advance to this day, still many of the New Synthesis tenets remain good science.

Consensus was the result of a number of factors. Perhaps most important was reconciliation between the Mendelians and the field biologists. The Mendelians, experimental geneticists who accepted Mendel's concept of particulate genetics, believed that evolution proceeds discontinuously through spontaneous, major gene (particulate units of heredity) mutations in a few individuals in populations, over one or a few generations. This is called mutation theory. Field biologists (naturalists) held an entirely different view of evolutionary change. In the tradition of Darwin, they believed that individuals within populations differ quantitatively or continuously from one another in heritable characters and that whole populations change gradually over time by selective processes working on this variation. In essence, naturalists believed that evolution, driven by natural selection, occurs slowly through gradual, small changes in continuously or quantitatively variable populations leading to greater adaptation of populations to their local environment. These two views differed so sharply from one another that initially they seemed irreconcilable.[12, 13, 15, 16, 17]

During the years after *Scopes*, the scientific landscape changed markedly as knowledge of genetics rapidly increased. Mutation theory began to erode as numerous genetics experiments showed that most heritable characteristics of organisms are not due to single genes, but to multiple genes acting together. The study of multiple gene inheritance became known as quantitative inheritance or polygenic inheritance. This provided a way to explain the continuous variation, that naturalists believed is seen in most populations in nature within accepted Mendelian genetic principles or laws. [12, 13, 15, 17] This reconciled two disparate fields within evolutionary biology by providing evidence that Mendelian genetic concepts are important in nature, but so is continuous variation within natural populations. In essence, the mutationists were exploring the correct genetic mechanisms, but they were incorrect that most evolutionary important change was the result of single gene mutations. The naturalists were correct that variation in natural populations is mostly continuous, and that entire populations transform over time, but it became understood that they do so through Mendelian genetic mechanisms.[12, 13, 15, 17]

Additional advances in genetics and population genetics also made the New Synthesis possible. Acceptance of August Weismann's view that genetic recombination and mutation are naturalistic sources of the variation that Darwin believed existed in populations and upon which natural selection works. This bolstered the view that natural selection is the main mechanism driving evolutionary change and undercut mutationist and other views of evolution. Darwin seemed to be vindicated[12, 13, 15, 17]

The major parts of the New Synthesis occurred between 1936 and 1950 as the conflict between the Mendelian geneticists and the naturalists was resolved and relative consensus was reached among biologists from diverse subdivisions of evolutionary biology. According to Mayr, the New Synthesis was a unification of the badly split fields of evolutionary biology. The result was the general acceptance of a number of concepts by evolutionary biologists from diverse disciplines. The ideas that recombination and mutation are the chief mechanisms for generating variation within populations were accepted. This supported Darwin's original belief that variation arises each generation by natural, nonsupernatural processes that are not environmentally directed.[12, 13]

The New Synthesis also accepted the Mendelian concept of particulate genetics (a theory of genetics that holds that the genetic units of the parents are particulate and they do not fuse or blend in the offspring but remain discrete over generations) ending what was a common belief in blending inheritance (the pre-Mendelian theory of genetics, which held that the genetic determinants of the parents fuse into a uniform substance during fertilization of the egg).[12, 13, 15, 17]

Also accepted was the idea that although mutations certainly occur, they usually have only small effects on organisms and do not generally result in significant evolutionary change in one or a few generations as was originally thought by the Mendelians. Furthermore, the understanding that most genetic characteristics are the result of multiple gene pairs acting together, rather than the action of only one pair, allowed evolutionists to accept the view that genetic characteristics are quantitative (continuous) in populations but result from Mendelian mechanisms of segregation and independent assortment.[12, 13, 15, 17]

Natural selection was accepted as the most important mechanism for evolutionary change, with the local environment seen as selecting for the new quantitative variations that occur every generation because of the naturalistic processes of mutation and recombination. In essence, the New Synthesis reaffirmed a Darwinian view that all adaptive evolutionary change is due to the directing force of natural selection acting upon abundant, random, intrapopulational variation.[12, 13, 15, 17]

Mayr asserted that two major conclusions emerged from the Synthesis.[12, 13, 18] First, evolution is gradual and capable of being explained in terms natural selection working on continuous variation produced by small mutations and genetic recombination. Second, all evolutionary phenomena can be explained in a way that is consistent with known genetic and population genetic mechanisms.[12, 13, 18]

As a result of the genetic advances leading to the New Synthesis and the acceptance of natural selection as the chief evolutionary mechanism, three major types of competing evolutionary theories (Lamarckism, orthogenesis, and mutation theory) were widely rejected by evolutionary biologists.[12, 13, 18]

The New Synthesis penetrated practically every field of biology, even paleontology. Paleontologists historically did not view evolution in terms of changing gene frequencies, but in terms of morphologic changes in organisms found in vertical, undisturbed, geological strata or sequences of rock. The presumption was that in undisturbed sequences the oldest layers are on the bottom and the youngest layers are on the top[12, 13, 18], allowing morphologic changes through time to be studied. In the midtwentieth century, the fossil record did not provide much evidence for Darwinian gradualism; rather, it generally seemed to support abrupt appearance of new morphological distinct forms. Despite these problems, in 1944 paleontologist George Gaylord Simpson attempted to reconcile a gradualist view of evolutionary change with the fossil record, which often seemed to indicate sudden appearance. This aspect of the fossil record troubled naturalists since before Darwin's *Origin*. Simpson, in his book, *Tempo and Mode of Evolution*, argued that such abrupt appearance of new species was apparent, rather than real, representing the fossil record's extreme imperfection and that it is not inconsistent with gradual sequential evolutionary change. This argument is similar to one made by Darwin in *Origin*, and it put paleontology within the framework of Darwinian gradualism and acceptance of the New Synthesis.[12, 13, 18]

The point of this story about the maturation of evolutionary biology is this. By the 1960s, widespread consensus prevailed in a post—New Synthesis scientific era. After the Synthesis, evolutionary biologists generally recast evolution mostly in terms of an updated Darwinian selection theory. Many evolutionary biologists at the time of *Epperson* took a view that evolution is a "fact" of nature and that the important mechanisms that drive evolutionary change had been solved by the Synthesis. All that remained in their opinion was working out the finer details of the evolutionary process.[12,13] It was an era where the scientific community had confidence that evolution was real and that its mechanisms were known. This optimistic view would be eroded somewhat in recent times, but much of the Synthesis still remains valid in many respects in the opinion of many evolutionary biologists.[19]

Analysis of the *Epperson* decision suggests that the optimistic view of most evolutionary biologists of that time was adopted by the court, and it became a determining factor in the decision.

Evolutionary biology was riding a wave of relative consensus, which the court bought into. However, the court did not explicitly acknowledge this and instead claimed that its decision to hold the Arkansas law unconstitutional was based entirely on the supposed sole religious purpose or motivation of the Arkansas legislature in enacting it. Yet a critical look at the facts shows this position to be disingenuous.

The *Epperson* decision specifically refers to evolution as "a particular segment" from the "body of knowledge." These phrases were directly derived from an *amici curia* brief by the National Education Association of the United States and the National Science Teachers Association, which was filed in this case on behalf of the plaintiff, Susan Epperson. As part of the brief, one hundred and seventy-nine biologists signed a statement maintaining that evolution has been "accepted into man's general body of knowledge by scientists and by reasonable persons who have familiarized themselves with the evidence." The court's specific reference to these phrases demonstrates the court's acceptance of evolutionary theory without real question or hesitation. It also suggests that the scientific status of evolution was a notable determinate in the court's decision. Further support for this assertion is seen in the court's statements that creationism is merely an anachronism that is indefensible in light of modern scientific knowledge.[3]

The court's acceptance of the scientific status of evolutionary theory explains its selective and biased use of the history of the law. It seems that its decision in the case was already made based upon its views of evolution and creationism, and the court used the legislative history selectively to support an already made decision. Serious criticism has been directed toward the court's holding that Fundamentalist religious fervor was the sole motivating factor behind the passage of the statute. Edward Larson, citing political scientist Virginia Gray, contended that the statute had broad-based support of two-third of the Arkansas voters, far more than can be attributed solely to Fundamentalist support. Although religious purpose unquestionably contributed to the passage of the law, other reasons were also important. For example, Bryan, in participating in the crusade that ultimately resulted in the enactment of both the

Tennessee and Arkansas laws, was equally concerned with preventing war and social exploitation. Bryan was convinced that these social ills followed unfailingly from Darwinism.[21, 22] Further, as Ronald Numbers pointed out, recent studies have shown that the people who joined Bryan's antievolution crusade came from all walks of life and all parts of the country.[20]

Although Larson has been challenged on this view, he has asserted that the *Epperson* court provided little documentation for its conclusions about the purpose of the Arkansas law. He contended that the only historical works cited by the *Epperson* court to support its conclusion were autobiographies by Darrow and Scopes, a 1937 ACLU report titled *The Gag on Teaching*, and a history of United States academic freedom by Hofstadter and Metzger. Larson claimed that these works constitute an obviously biased sample of the available literature.[21, 22]

Larson also criticized the court's use of the *Arkansas Gazette* advertisement as its best direct evidence for sole religious purpose. He stated that the advertisement was obscure and appeared only twice in Little Rock newspapers in the two-week period preceding the popular vote on the statute. In addition, Larson aimed a serious charge against the author of the *Epperson* decision (Justice Fortas), accusing him of altering the meaning of the advertisement by using ellipses to delete two central sentences from the portion of the ad quoted in the court decision. According to Larson, the edited version of the advertisement substantially altered its meaning. For example, the original ad stated, "The bill does not prohibit free speech, it does not seek to help the church. It simply forbids the state attacking the church by having evolution taught in the schools at taxpayer's expense." Larson contended that the unedited version of the ad does not support a sole purpose interpretation; rather, it supports Bryan's stated purpose of promoting religious neutrality. Bryan assumed that creationism could not be legally taught in public schools, and therefore he sought neutrality by also removing evolution. According to Larson, despite these well-known facts, Fortas purposely deleted contradictory sentences from the portion of the ad quoted in the *Epperson* decision to make it appear that Law cannot be defended as an act of religious neutrality, but solely as an effort to promote creationism.[21, 22]

Nothing in this discussion is meant to deny that religious motivation was behind the enactment of the Arkansas law, or to support any creationist claim against evolutionary theory. Ample evidence for religious motivation is found from the express language of the law, as well as significant Fundamentalist religious involvement in its enactment, and I do not support a creationist interpretation of the history of life or earth history. Notwithstanding this, the *Epperson* decision is flawed and wanting in the fundamental fairness that we have a right to expect from our courts and government. The court's purpose analysis of the law is tainted by its own biased notion of what constitutes scientific legitimacy and by its own bias that the Arkansas law represented an attack on a major scientific theory. The *Epperson* decision suggests that the Supreme Court accepted the view of the mainline scientific community, without real analysis, and then used the *Schempp* purpose test to support its preconceived notion. The *Epperson* decision is a classic example of the right decision for the wrong reasons. This criticism will appear again in this work in later discussions concerning subsequent evolution in public school cases. *Epperson* would have been on a much stronger legal and ethical foundation to rule that the predominate purpose and effect of the Arkansas Stature was religious in nature, which is sufficient to invalidate it, and to give judicial notice that under no reasonable or generally understood definition can evolution be legally defined as religious.

In any event, *Epperson* did not end legal controversy concerning evolution in public schools; it only marked the beginning of federal court involvement. A variety of alternative legal paths were subsequently tested by those interested in challenging the place of evolutionary theory in public schools. Nonetheless, after *Epperson*, every federal court that decided a public-school evolution case did so using First Amendment criteria. Since *Epperson*, all antievolution statutes or state actions designed to limit the teaching of evolution in the public schools have been cast as First Amendment religious-clause matters.

FEDERAL EVOLUTION LITIGATION IN THE 1970S: THE FIGHT TAKES NEW DIRECTIONS WITH THE ORIGIN OF EQUAL TIME CHALLENGES

Wright v. Houston Independent School District and Daniel v. Walters

The *Epperson* decision did not end legal disputes concerning evolution in public schools. It actually turned out to be a beginning. Federal cases of this type have continued into the twenty-first century. Controversy has not abated to this day. In the 1970s, opposition centered on the exclusivity afforded evolutionary explanations in mainstream biology and science education; creationist sought equal time in the public-school science curriculum with evolution.

The exclusivity afforded evolutionary explanations reflects its scientific status. The "fact" of evolution has been a settled issue in mainstream biology since before the turn of the twentieth century. Scientific challenges to evolutionary explanations of life history have been mostly absent in biology. Creationist explanations have largely been ignored. Creationism has, explicitly and implicitly, been denied any semblance of scientific legitimacy. Therefore, it is not surprising that in the litigious environment of late-twentieth century and early twenty-first century United States, certain types of creationists would turn to the federal courts to achieve the credibility that has otherwise been denied them in the laboratories, journals, and classrooms.

The 1970s and 1980s were marked by a number of evolution in public-school cases litigated in the federal system. The creationist challenges to evolution in the public-school curricula relied on both contrived scientific arguments and sophisticated legal arguments to support them. Notably, scientific creationists have attempted to devalue the significance of mainstream science's almost unanimous support for evolutionary explanations and mainstream science's almost total rejection of creationism. Creationists focused upon and emphasized disagreement among evolutionists concerning evolutionary mechanisms. Please allow me to explain.

New Synthesis Darwinism during this time period was being criticized within some scientific circles for various reasons. The most important had to do with what mechanisms drive evolutionary change, and the mode and tempo of such change. To be sure, there was virtually no dispute as to whether evolution occurred or is occurring. Evolution remained a biological fact to scientists. Notwithstanding this, creationists focused on the disagreement among evolutionists concerning mechanisms to bolster the legitimacy of their views on life history. For example, an important creationist challenge to evolution was derived from the work of philosopher of science, Karl Popper. Popper at one point in a nuanced way rejected the scientific validity of Darwinian evolutionary theory. Another challenge to evolution was derived by co-opting the ideas of paleontologists Stephen Jay Gould and Niles Eldredge concerning the mode and tempo of evolutionary change, called punctuated equilibrium. These issues will be more completely discussed in subsequent chapters.

There was an important legal aspect to these creationist challenges. They had to be structured within the First Amendment Establishment Clause and Free Exercise of Religion Clause constitutional framework. The legal precedent set by *Epperson* mandated this. Because of *Epperson,* most politically active creationists understood the futility of attempting to directly ban evolutionary theory from public school curricula. So they formulated alternative strategies. They attempted to show that evolution is so lacking as a scientific theory that an alternative scientific creationist interpretation, with its emphasis on the uniqueness of the human species, abrupt appearance of species in the fossil record, and a young earth, is just as plausible, if not more so, and therefore should be recognized by the courts as a legitimate alternative to evolution in the public-school science class-rooms and be given equal time.[1]

An important creationist strategy, used in the decade of the 1980s to achieve their goals, was enactment of "balanced treatment" statutes mandating that creationism be given equal treatment with evolution in the science curriculum. Balanced treatment will be discussed in later chapters of this work.[2, 3, 4]

Here, we discuss two salient federal evolution cases from the 1970s, *Wright v. Houston Independent School District* and *Daniel v. Waters*. These cases illustrate the scope of the antievolution judicial and legislative challenges during that decade as creationists continued to struggle for legitimacy of their ideas in public science education. The *Wright* and *Daniel* cases represent two different ways creationists attempted to achieve equal time goals. Equal time became the legal and intellectual parent of the balanced treatment statutes of the 1980s.

Wright v. Houston Independent School District

The first federal case after *Epperson* to deal with the issue of evolution in the public-school science curriculum was *Wright v. Houston Independent School District*[5] in 1972. In *Wright*, the plaintiffs, a number of students in Houston, Texas, initiated an action in federal district court seeking to stop the school district and State Board of Education from including the theory of evolution as part of the academic curriculum and from adopting textbooks that presented evolutionary theory without critical analysis and without the inclusion of other theories regarding the "origin of man;" namely, those derived from the Bible.[5] Plaintiffs alleged that the presentation of evolution in the public-school curriculum without such constraints inhibited the free exercise of their religion and constituted an establishment of religion by the state in violation of both religious clauses of the First Amendment to the federal constitution. In support of these assertions, the plaintiffs alleged that the theory of evolution was so antagonistic to the creation account in *Genesis* that it constituted a direct attack by the state upon their religious beliefs. Plaintiffs incredibly attempted to use the *Epperson* decision as legal precedent to limit the teaching of evolution in public schools.

They argued that evolution was really part of the religion of secular humanism in the guise of a scientific theory. Therefore, the plaintiffs alleged that the method of presentation of evolution in the Houston School District violated constitutional principles of government neutrality in matters of religion as articulated in *Epperson*.[5]

Rejecting the plaintiffs' argument, the district court dismissed this case, on a motion by defendants without a trial, holding that the plaintiffs failed to state a legally valid claim or establish any valid analogy to *Epperson*. The court wrote that, whereas the Arkansas law, litigated in *Epperson*, made the mere reference to evolution in public schools a criminal offense, the State of Texas and the Houston School District did not enact any legislation or directive supporting any view on the subject. The *Wright* court stated that the defendants, at most, had a general policy of approving textbooks that presented the theory of evolution in a positive light. In addition, the *Wright* court stated that the defendants did nothing to discourage any discussion of human origins or deny students the right or opportunity to challenge their teachers' presentation of evolutionary theory.[5]

Wright rejected the plaintiffs' claims that the First Amendment religious clauses were violated. The court refused to expand any current legal definition of religion as including evolution. Although nowhere in *Wright* did the court attempt to define religion, the court contended that there was no factual support for defining evolution as religion as that term is used in First Amendment litigation. The court asserted that just because science and religion often deal with some of the same questions and frequently reach conflicting answers, it is not the business of the court to intervene by suppressing real or imagined attacks on a particular religious doctrine. The court stated that teachers of science in public schools should not be expected to avoid discussion of scientific issues because certain religions disagree with them.[5]

Furthermore, the court also rejected the plaintiffs' request that the school district give "equal time" to all theories regarding human origins. The court stated that if the beliefs of fundamentalism were the sole alternative to evolution, such remedy might be feasible. However, the court asserted that every religion holds its own particular view of human origins, and even within the scientific community itself there is much debate over the details or mechanisms of evolution. Therefore, the court concluded that it is impossible to insist upon the presentation of all theories of human origins.[5]

This is actually an important issue. If a court mandates equal time for opposing views of how humans got here, where does it end? Every culture, every society, every religion has its own view, oftentimes unique to it. Why would the constitution only deem a Fundamentalist or evangelical, hyperliteral, Judeo-Christian view of origins as the only alternative to evolution worthy of legal protection? Or in the alternative, how does a court prohibit evolutionary theory, embraced by almost the entire scientific community, from being taught?

In considering these problems, the *Wright* court rejected the assertion that evolutionary theory is a religion as that term is broadly understood under the federal constitution and First Amendment litigation.[5] It also declined to interpret *Epperson* as precluding the teaching of evolution unless other theories, namely creationism, were also included.

The *Wright* decision showed some of the same analytical problems as *Epperson;* it presumed that evolutionary theory is a scientific theory, while concluding that creationism was thinly veiled religion. Once again, we see a federal court adopt carte blanche mainstream science's view on this matter without any analysis or justification. However, there are analytical differences between *Wright* and *Epperson* that make *Wright* much easier to justify. No strained analysis by the court was made in *Wright* such as we see in *Epperson.* A creationist motive and effect to brand evolution religious and to mandate equal time for creationism was palpable. Courts have a right to take judicial notice of the obvious, and two things were obvious in *Wright*: evolution is not religion by any reasonable and generally understood definition of that term, and the motive and effect of the plaintiffs' position were obviously religious. The court felt that these interpretations were so correct that no legal definition of religion needed to be given. Judges do not live in a vacuum but within a society. *Epperson* would have been well advised to take the more constrained position similar to *Wright* in its analysis.

Wright was appealed to the Court of Appeals; however, the Appeals Court affirmed the lower-court decision dismissing the suit, and the Supreme Court of the United States refused to hear the case.[5]

In summary, *Wright* rejected the plaintiffs' request for equal time. This represents the first equal-time claim brought before a federal court. Although it was defeated in *Wright*, equal-time statutes became the next important challenge to face evolution in public schools.

Daniel v. Waters: The Beginnings of Equal-Time Statutes

An equal-time challenge was presented in a Tennessee federal court in the case *Daniel v. Waters*.[6] This case, decided in 1975, is significant because it represents the first time an "equal time" statute was tested.

After *Epperson* and *Wright*, it became clear to creationists that removing evolution from the public-school curriculum would be unlikely for First Amendment constitutional reasons. *Daniel* represents a creationist response to the *Epperson* and *Wright* decisions, attempting to get creationism equal time with evolution in public-school science classes through equal time statutes. Equal-time statutes, called balanced treatment statutes in the 1980s and beyond, became the basis of subsequent evolution litigation in the federal courts.

The *Daniel* case resulted from the passage of an anti-evolution statute in 1973 by the Tennessee legislature. The statute required "Any biology textbook used for teaching in public schools, which expresses an opinion of, or relates a theory of the origins or creation of man and his world shall be prohibited from being used as a textbook in such system unless it specifically states that it is a theory as to the origin and creation of man and his world and is not represented to be a scientific fact. Any textbook so used in the public education system which expresses an opinion or relates to a theory or theories shall give in the samse textbook and under the same subject commensurate attention to, and an equal amount of emphasis on, the origins and creation of man and his world as the same is recorded in other theories including, but not limited to, the *Genesis* account in the Bible... The teaching of all occult or satanical beliefs of human origin is expressly excluded from this Act... Provided, however that the Holy Bible shall not be defined as a textbook, but is hereby declared to be a reference work and shall not be required to carry the disclaimer above provided for textbooks."[6,7]

The individuals who brought the action were teachers of biology in the Tennessee public school system and a scientific organization, the National Association of Biology Teachers. The defendants included members of the Tennessee State Board responsible for selecting public-school textbooks. After a rather complicated judicial history, during which a State Chancery Court of Davidson County, Tennessee, held that the Statute in question was in violation of the First and Fourteenth Amendments of the federal constitution, the case ultimately found its way to the Federal Appeals Court, 6th Circuit, Tennessee.[6]

The Federal Appeals Court held that the statute litigated in *Daniel* was unconstitutional on its face as being in violation of the First Amendment to the federal constitutiosn. Citing *Epperson*, the *Daniel* court stated that the First Amendment prohibited any law "respecting the establishment of religion." Applying the *Epperson* standard, the court concluded that the statute in question provided preferential treatment of the Bible and violated the First Amendment for at least four reasons. First, the statute prohibited the selection of any textbook that teaches evolution unless it also contained a disclaimer that evolution was only a theory as to the "origin and creation of man and his world and is not to be represented as scientific fact." Second, the statute required the inclusion of the *Genesis* version of creation, if any version is taught at all, while permitting the *Genesis* version to be treated without a disclaimer. Third, the Bible under the statute was exempt from being treated as a textbook, and it was not required to carry the disclaimer. Fourth, the statute prohibited "satanical" or "occult" beliefs of human origins from being represented in textbooks.[6]

The religious nature of the statute litigated in *Daniel* is beyond question. This decision not only marks the first time an equal-time statute was reviewed by a federal court, but perhaps even more importantly, it marks the first time a court used the *Lemon* test to determine constitutionality of an antievolution statute. In doing so, the *Daniel* court held that the portion of the statute prohibiting the teaching of all "occult" or "satanical" beliefs of human origin from inclusion in textbooks would involve the Tennessee State Textbook Commission in difficult theological arguments in direct conflict with the third prong of the *Lemon* test establishment-clause analysis. The third prong requires that a statute must not foster an excessive government entanglement with religion.[8]

The *Lemon* test, as a test for First Amendment Establishment Clause violations, was set forth by the Supreme Court of the United States in *Lemon v. Kurtzman*[8] in 1971. In *Lemon*, the Supreme Court extended an existing First Amendment test, first articulated in *Abington School District v. Schempp* in 1963[9] and reaffirmed in *Epperson*. The court in *Lemon* proposed three criteria to test a statute for Establishment Clause violation. These three criteria became the so-called three prongs of the *Lemon* test. First, the statute in question must have a secular purpose. Second, its principal or primary effect must be one that it neither advances nor inhibits religion. Third, the statute must not foster an excessive entanglement with religion. A statute violates the Establishment Clause of the First Amendment if it fails to satisfy even one of these three prongs.

As a result of the blatantly religious nature of the Tennessee Statute and the prior development sof First Amendment constitutional law in *Epperson* and *Lemon v. Kurtzman*, the *Daniel* court invalidated the statute as violating the Establishment Clause of the First Amendment to the federal constitution. The statute failed to meet the requirements of all three prongs of the *Lemon* test.

Most likely, because of the obvious religious nature of the statute in question, the *Daniel* Court did not feel it needed to address the scientific status of evolutionary theory as *Epperson* and *Wright* did.

Daniel v. Waters is a sound, judicially constrained decision, legally grounded in the law of that time. It did not contain the judicial overreach that will be seen in subsequent evolution cases that will be discussed in the chapters to come.

The *Wright* and *Daniel* decisions sent a strong signal to creationists that successful federal litigation would not be a smooth road. However, these lower federal court rulings did not deter creationist endeavors of this type; rather, they helped shape its form in the decade of the 1980s.

FEDERAL EVOLUTION LITIGATION IN THE 1980S: AN ARKANSAS BALANCED TREATMENT STATUTE

McLean v. Arkansas

Equal time for the teaching of creationism in the public schools was codified by "balanced treatment" statutes in the 1980s. These statutes, in one form or another, characterized federal evolution litigation in that decade and beyond. An Arkansas statute of this type was litigated in one of the most important evolution cases ever decided in the federal court system, *McLean v. The Arkansas Board of Education.*[1] *McLean* was a 1982 Arkansas federal district court case. In the 1980s, balanced-treatment legislation was written and promulgated by groups espousing a type of creationism known as creation science, or as it is sometimes called, scientific creationism.

In March, 1981 the "Balanced Treatment for Creation-Science and Evolution-Science Act"[2] was enacted into law in Arkansas. It provided that public schools within the state "give balanced treatment to creation-science and to evolution-science." The act contained a specific definition of both. We will be discussing what creation science is a bit later in this chapter in great detail, including its tenets, history, and importance to the evolution in public-school cases. But for now, let's consider how the act defined this body of knowledge.

The act defined creation science as including "the scientific evidences and related inferences that indicate sudden creation of the universe, energy, and life from nothing; the insufficiency of mutation and natural selection in bringing about the development of all living kinds from a single organism; changes only within fixed limits of originally created kinds of plants and animals; separate ancestry for man and apes; explanation of the earth's geology by catastrophism, including the occurrence of a worldwide flood; and a relatively recent inception of the earth and living kinds."[2] This definition left no doubt in any reasonable person's mind who looked into this matter in a fair way that it was an direct and unambiguous restatement of what was called "creation science."

The act defined "evolution science" as including "the scientific evidences and related inferences that indicate: emergence by naturalistic processes of the universe from disordered matter and emergence of life from non-life; the sufficiency of mutation and natural selection in bringing about development of present living kinds from simple earlier kinds; emergence by mutation and natural selection of present living kinds from simple earlier kinds; emergence of Man from a common ancestor with apes; explanation of earth's geology and the evolutionary sequence by uniformitarianism; and an inception several billion years ago of the earth and somewhat later of life." The definition of evolution was basically a restatement of many, not all, New Synthesis ideas but was scientifically incomplete for the time period that the case was decided. The obvious failure of this definition is that it did not take into account how evolutionary science had progressed during the 1970s. As will be discussed, this failure became an important issue in the trial and the ultimate decision of the court in this case.

A lawsuit was filed in Arkansas federal district court challenging the validity of the act on grounds that it violated the Establishment and Free Speech clauses of the First Amendment to the federal constitution. The plaintiffs in this case included an eclectic group including bishops of various Christian denominations, clergy of various denominations, parents, at least one biology teacher, Jewish organizations, the National Association of Biology Teachers, and other taxpayers and religious and political organizations. The defendants included the Arkansas Board of Education, the director of the Department of Education, and the State Textbook and Instructional Materials Selecting Committee. A trial was held in December 1981; the named plaintiff was a clergyperson from a mainstream denomination.[1] From this list of plaintiffs, we can see an alignment of mainstream Western religions with mainstream scientific organizations. It is no coincidence that the lead plaintiff was a clergyperson; this case from the beginning was simply not a matter of religion versus science. It never was that, no matter how some try to portray it.

The *McLean* court, as all federal courts since *Epperson*, decided this case using First Amendment, Establishment Clause criteria. This required application to the Balanced Treatment Act the prevailing test during that time period for Establishment Clause constitutionality, the *Lemon* test. In applying *Lemon*, the *McLean* court turned to the first prong of the test, the purpose prong; that is, a statute must have a secular legislative purpose. In doing so, the court assessed the legislative purpose of the act by reviewing its history.[1]

Because the act mandated balanced treatment of creation science with evolution in public schools, the court began its purpose analysis by trying to determine what creation science really is. Based upon testimony of various expert witnesses, the court stated that creation science emerged around 1965 from Fundamentalist religious roots. In the 1960s and 1970s, from these roots, organizations such as the Creation Research Society and the Institute for Creation Research were formed to promote the idea that the *Genesis* account of creation can be supported by scientific data. The terms *creation science* and *scientific creationism* were adopted by these organizations to describe their work. The court stated that these organizations and modern creationists adopted a Fundamentalist view that there are only two positions with respect to how living organisms got here: a belief in the historical accuracy of the *Genesis* story of creation or a belief in evolution. These Fundamentalist organizations considered the introduction of creation science into the public schools' part of their religious ministry, and they viewed evolution as a source of society's ills.[1] The court concluded, quite reasonably, from this history that creation science is a front for Fundamentalist religious belief.

The *McLean* court also focused on the legislative history of the Balanced Treatment Act. As part of this, it explored the activities of the statute's draftsperson, Paul Ellswanger. The court noted that Ellswanger was the founder of an organization called Citizens for

Fairness in Education. As a respiratory therapist by education and profession, the court concluded that he was untrained in law and science. The court reviewed his deposition testimony and his various correspondences, concluding that Ellswanger's endeavors were religious in nature and that he attempted to conceal that fact. The court believed that his motivation was opposition to evolution, the "idea of killing evolution," in Ellswanger's words, and a desire to see a literal bible version of creation taught in public schools.[1] The court supported this by quoting parts of two Ellswanger's letters to the act's supporters. One to Louisiana State Senator, Bill Keith, where Ellswanger wrote, "I view this whole battle as one between God and Anti-God forces," and "It does the bill effort no good to have ministers out there in the public forum and the adversary will surely pick up on this point." The court also quoted Ellswanger to a Florida State senator, stating, "all of us engaged in this legislative effort be careful not to present our position and our work in a religious framework."

The court also considered the activity of Senator Holstead's in support of the act. Holstead was an Arkansas legislative sponsor of the act. According to the court, Senator Holstead departed from normal legislative procedures during his sponsorship. The court pointed out that he failed to consult with appropriate state agencies or scientists before the act was introduced into the Arkansas legislature, and no representative of the State Department of Education testified in the legislature during hearings on the act. The court emphasized that Senator Holstead himself testified that his sponsorship and lobbying efforts on behalf of the act were motivated by his religious beliefs and his desire to see the biblical version of creation taught in the public schools.[1]

The court asserted that the act was passed after only a few minutes' discussion on the State Senate Floor. In the State House of Representatives, the act was referred to the Education Committee, which the court stated conducted only a fifteen-minute hearing with no scientist or representative from the State Department of Education testifying. In addition, the court related that the act was enacted into law without amendment or modification, and no meaningful fact-finding process was employed by the general assembly.[1]

Finally, the court felt that Arkansas's long history of official opposition to evolution, motivated by religion, could not be ignored. As we have seen from previous chapters, it dates back to the 1928 antievolution statute that led to the *Epperson* decision in 1968.[1]

After considering all of this, the *McLean* court held that the act was enacted with the specific purpose of advancing Fundamentalist religion, thereby failing the purpose prong of the *Lemon* test.[1] After determining this, the *McLean* court did not need to consider the matter any further. Remember, failure to pass even one prong is constitutionally fatal. However, the *McLean* court was determined to put the nails in the coffin of creation science as a viable alternative to evolution in the public-school science curriculum.

The McLean court next turned to the second prong of the *Lemon* test, the effect prong; that is the principal or primary effect of a statute must be one that neither advances or inhibits religion. The court believed that the Balanced Treatment Act failed here also. The court concluded that the act served merely to ratify a Fundamentalist religious interpretation of *Genesis*; as such, the act's major effect was the advancement of religion. The court stated that the two-model approach to teaching creation science and evolution science found in the act is identical to the two-model approach espoused by the Institute for Creation Research, a Fundamentalist organization, and it was taken almost verbatim from its writings. The court believed that this approach was an extension of Fundamentalist view that one either accepts a literal interpretation of *Genesis* or else believes in the godless system of evolution.

Once again, the court reasonably accepted the testimony of plaintiff witnesses who asserted that there are a number of theories about the origin of life in addition to evolution and Fundamentalist creationism. The court stated that the two-model approach was a "contrived dualism," which has no scientific basis or legitimate educational purpose; that it merely casts in educationalist language the dualism that appears in creationist literature.[1]

The court also criticized the methodology used in creation science, stating that creationist-scientists do not collect data, weigh it against opposing scientific data, and thereafter reach conclusions. Rather, the court stated that creationists take the literal wording of *Genesis* and attempt to find scientific support for it. Further, the court asserted that proof in support of creation science consists almost entirely of efforts to discredit evolution, rather than independent laboratory research or investigation. The court stated that no scientific journal has published a paper supporting creation science as defined in the act, nor did any creationist witness at trial produce an article that had been submitted for publication. The court stated that the defendants and creationists in general believe that any evidence that fails to support evolution necessarily supports creationism and is creation science. The court concluded that creation science failed to meet the requirements of a scientific theory and failed to meet the definition of science.[1] The court asserted that the only real effect of the act was the advancement of religion. Therefore, the court held that the act violated the second prong of the *Lemon* test as well as the first prong.[1, 3]

The court also concluded that the act failed the third prong of the *Lemon*, the entanglement prong; that is, a statute must not foster an excessive government entanglement with religion. The court stated that evolutionary thought permeates public-school textbooks and teaching through a variety of intellectual disciplines, not just biology. Other subjects, such as world history, geology, zoology, botany, psychology, anthropology, sociology, philosophy, physics, and chemistry all contain evolutionary ideas. The court felt that implementation of the act would require public schools to not teach significant portions of these subjects, or to balance them with scientific creationist views, which the court believed was religion. The court asserted that balancing would require the state to constantly monitor educational materials. As a result, the court concluded that state entanglement with religion is inevitable under the act, in violation of the third prong of the *Lemon* test.[1, 3]

While engaging in its *Lemon* analysis of the act, the *McLean* court needlessly entered into a complicated and slippery area of philosophy by attempting to construct a judicially workable definition of science. The court stated that creation science failed to meet the essential characteristics of science. The court asserted that to be science an activity must be guided and explained by natural law, testable against the empirical world with tentative and falsifiable conclusions[1]

The court applied its definition of science to the various sections of the act and found them wanting in this regard. For example, the court stated that the section of the act defining creation science as sudden creation "from nothing" is not science because it relies on supernatural intervention, which is not guided by natural law and which is not testable or falsifiable. The court rejected the defendants' assertion that the concept that creation from nothing does not necessarily involve a supernatural deity. Rather, the court concluded that creation out of nothing is a concept unique to Western religions, and the concept of a creator of the world is equivalent to a concept of God. The court further refused to accept as constitutional any defense argument that teaching the existence of God is not religious if it doesn't seek a commitment from the believer.[1]

The court stated that the section of the act that asserts that mutation and natural selection are insufficient to bring about the existence of present living kinds from simpler kinds is nothing more than an incomplete, negative generalization and not legitimate scientific analysis. In addition, the court felt that the section that asserts that organisms can change over time "…only within fixed limits of originally created kinds…" also fails to conform to the essential characteristics of science since there is no scientific definition of "kinds," and there is no scientific explanation for "fixed limits" or boundaries of change.

The court continued its scathing criticism of the act, opining that the section on the "separate ancestry of man and apes" is a statement that explains nothing. The section that refers to the "explanation of the earth's geology by catastrophism, including the occurrence of a worldwide flood," fails as science because a worldwide flood as an explanation of the world's geology cannot be explained by natural law. The section on the "relative recent inception" of the earth, according to the court, has no scientific meaning and is calculated in reference to Old Testament genealogies, rather than explainable by natural law; further, the court believed it is not tentative.[1]

47

The court next turned its attention to the act's definition of evolution, criticizing it as factually limited, incomplete, and incorrect. The court quoted the act, "the sufficiency of mutation and natural selection in bringing about the existence of present living kinds from simpler kinds" through evolutionary change. Two plaintiff scientific experts on evolution, Francisco Ayala and Stephen Jay Gould, contradicted this, testifying that biologists know that these two processes do not account for all significant evolutionary change. They testified that such phenomena as recombination, genetic drift, and the theory of punctuated equilibrium, are also believed to play an important role in the evolutionary process.

Punctuated equilibrium was an especially important theory at the time of *McLean and Edwards v. Aguillard* (discussed next chapter). It not only impacted on how scientists view the tempo and mode of evolution change, but directly undercut creationist "evidence" for divine creation. In 1972, paleontologists, Stephen Jay Gould and Niles Eldredge, published a paper on what they called punctuated equilibrium.[4] The paper was a response to New Synthesis emphasis on gradualism; that is, the concept, adopted from Darwin, that evolution generally is the result of a gradual accumulation of small changes over long periods of time. New Synthesis evolutionists believe that both microevolution (defined as fluctuations in gene frequencies from generation to generation) and macroevolution (the process of speciation and even emergence of higher taxa) proceed gradually, basically through the same mechanisms. Microevolution and macroevolution were distinguished only by the length of time required to produce the change, and of course the extent of the change. Essentially, this view resulted in a deemphasis of macroevolution by viewing it merely as an extension of gradual microevolution. In 1944, Simpson, one of the architects of the New Synthesis, included paleontology into the Synthesis by interpreting the fossil record's evidence of evolution in a way that did not contradict gradualism. His interpretation was not obvious, since a careful look at fossil evidence at that time generally suggested sudden, nongradual appearance of new species rather than gradual emergence. Naturalists, since before Darwin's *Origin of Species*, noticed this saltational data from the fossil record. Certainly, Darwin noticed it. Simpson, in his book *Tempo and Mode of Evolution*, argued that such abrupt appearance of new species was apparent and artefactual rather than real. He claimed that it represents the fossil record's extreme imperfection, and it is not a repudiation of gradual sequential evolutionary change. This argument is similar to one made by Darwin in *Origin*, and it had the important effect of placing paleontology within the framework of Darwinian gradualism.[4]

By the early 1970s, Gould and Eldredge, not satisfied with this long-standing explanation, proposed punctuated equilibrium. In doing so, they hoped to validate the fossil record as a source of valuable positive data rather than an extremely imperfect manifestation of gradualism as Darwin, Simpson, and others had characterized it. Most paleontologists during the 1970s and 1980s were familiar with the fact that despite the New Synthesis interpretation, and much new fossil data collected since the 1940s, the fossil record of many species shows little gradualistic change. Many species are characterized by a geologically unobservable origin, or apparent sudden appearance, followed by morphological stability, or stasis, until extinction.[4] Gould and Eldredge attempted to explain the fossil record by application of an idea first expounded by Ernst Mayr in the 1950s. Mayr proposed that most speciation occurs by selection in small populations cut off in remote corners at the edge of the original species range (called peripheral isolates). This type of speciation is known as peripatric speciation, which is a special type of allopatric speciation (speciation occurring in geographically isolated populations). Peripatric speciation is characterized by the small size of the separated population. Although Mayr's idea remained Darwinian because of its continued reliance on gradual change and natural selection, it did differ from the commonly held New Synthesis views that most evolution occurs within large populations of organisms, or sympatrically as it is termed. However, Mayr contended that although speciation was gradual in peripheral isolates, it is often "rapid" in geological terms (for example it might only take a few thousand years). Gould and Eldredge suggested that events of this type are unlikely to leave any trace in the fossil record because of the relative "rapidity" of the process, geologically speaking, and the very small size and area which the changing population initially existed. Therefore, according to a punctuated view, the fossil record need not be explained by scientists as a hopelessly imperfect fraction of a gradual gradation of evolutionary change within large populations, but an actual and accurate reflection of evolutionary events that occur[4] According to Gould and Eldredge, occasionally after speciation has taken place within a peripheral isolate, the new species may reenter the area occupied by the parent species. If the new species is better adapted to the local environmental conditions than the parent species, it potentially can outcompete the parent species, eventually spreading over a wide area. As the new species' range increases, the chances of fossilization increase concomitantly. Under these circumstances, the new species will seem to appear "suddenly" in the fossil record, giving the appearance of a totally new form with no continuous links back to the parental form.[4]

A second characteristic of punctuated evolution is stasis or failure of a species to significantly change in form after speciation. The failure of species to change appreciably after appearance was another characteristic paleontologists frequently noticed. Under a punctuated view, after speciation, new species are thought morphologically fixed in character, and they remain largely unchanged until extinction, apparently because of genetic constraints.

This view of speciation engendered considerable controversy especially at the time *McLean* was being litigated. As an aside, evolutionist Douglas Futuyma has recently reviewed numerous fossil lines of different species for evidence of gradual or punctuated modes of change. He conceded that the fossil record provides evidence for both patterns: neither one nor the other solely reflects the record.[5]

Not only did punctuated equilibrium explain what scientists have historically observed about the fossil record, but it also directly undercut creationist interpretations of the fossil record as supporting a sudden appearance of new species only, or best explained by special creation by a creator.

In addition to all of this, the court also criticized how the act dealt with the topic of origin of life. The court asserted that scientific creationist views on this issue are basically incorporated into the act. The court stated that although the subject of the origin of life is within the province of biology, the scientific community does not consider origin of life a part of evolutionary theory. Rather, the court concluded, the theory of evolution grants that life exists and instead focuses on how it changes over time. Therefore, the court stated, the act's inference that evolution presupposes the absence of a creator is incorrect. In support, the court further cited Ayala, a geneticist and theologian, who pointed out that many working scientists who subscribe to the theory of evolution are devoutly religious.[1]

The *McLean* decision is controversial. On one hand, it has been hailed by *Science*, the leading journal of American science, as "the finest legal document ever written about this question—far surpassing anything that the *Scopes* Trial generated, or any document arising from the two Supreme Court cases... Judge Overton's definitions of science are so cogent and clearly expressed that we can use his words as a model for our own proceedings." *Science* published Judge Overton's opinion verbatim as a major article.[6] At the other extreme, *McLean* has been criticized as representing an overbroad opinion that overestimates the open-mindedness of scientists and ignores the dangers inherent in a federal court strictly defining subjects within the public-school curricula[7], especially subjects that it has no expertise or legitimate ability to address. I lean more toward the latter interpretation, although I believe creation-science has no place in the public school science curriculum.

A critical review of the *McLean* decision reveals several troubling issues. First, the *McLean* court ignored legal precedent of that time in reaching its decision. Although few would disagree with the court opinion that the history and legislative history of the Arkansas statute established a clear religious motive or purpose, the court ignored a key question in arriving at that decision. That is, was religion the sole purpose of the act? This was not trivial legal question. At the time *McLean* was decided in early 1982, the Supreme Court of the United States had invalidated a state statute for lack of secular purpose (what has become the first prong of the *Lemon* test) only twice, in the *Epperson* evolution case in 1968 and in the case, *Stone v. Graham*, in 1980.[8, 9] In *Stone*, the Supreme Court ruled that a Kentucky law that required the posting of the Ten Commandments on the wall of every public school classroom in the state violated the Establishment Clause of the First Amendment because the purpose of the display was essentially religious. In both cases, the Supreme Court found that the respective statutes were solely or wholly motivated by religious purpose, an admittedly weighty standard of review. However, despite this precedent, the *McLean* court engaged in no meaningful analysis of whether a finding of sole religious purpose was necessary to invalidate any statute, including the Arkansas Act, for violating the first prong of the *Lemon* test. Rather, the court merely concluded that the correct standard of review is a specific purpose or primary purpose test, with no discussion of the relationship of this test to a sole purpose test.[10, 11, 12, 13] This issue would be addressed in the evolution and public school Supreme Court case discussed in the next chapter, *Edwards v. Aguillard*.

A second analytic problem with the *McLean* decision was the court's conclusions on the legislative motive in enacting the act. The court concluded that the expressed motives of the draftsperson and sponsors of the act (some of whom were not even from Arkansas) were sufficient to support the conclusion that the primary or specific purpose of the Arkansas legislature in passing the act was religious. The court also concluded that the history and legislative history of the act was sufficient to support the conclusion that no one in the state legislature considered the educational value of the act. However, there are logical and analytic problems associated with imputing universal motives on a group from the actions and comments of a few members, even when those few have a leadership capacity.[10, 11, 12, 13]

The *McLean* court was on firmer ground when it concluded that the act had a religious purpose after reviewing the act's definition of creation science and comparing it with a Fundamentalist interpretation of the *Genesis* account of creation. This provided strong evidence for imputing legislative purpose and motive.[10, 11, 12, 13]

As stated, since all three prongs must survive *Lemon* scrutiny, the *McLean* court reasonably concluded that the statute failed the *Lemon* test and, as such, violated the Establishment Clause of the

First Amendment. The court did not need to continue its analysis. Notwithstanding this, the *McLean* court, in what could be considered an absence of judicial restraint, unnecessarily entered into a detailed and controversial discussion of what science is and what it is not.[10, 11, 12, 13]

The court allowed the testimony of Plaintiffs' expert Michael Ruse, a philosopher of science to instruct the court on the nature of science. Ruse relied on philosopher Karl Popper's method of demarcating science from nonscience in his testimony. Accepting Ruse's arguments, the *McLean* court determined that the essential characteristics of science include naturalness, tentativeness, testability, and falsifiability. The court concluded that creation science failed to meet these criteria, and as a result it is not science but religion.[10, 11, 12, 13]

However, in doing this, the *McLean* court oversimplified a complicated demarcation of science/nonscience argument in a way that was disingenuous at best and deliberately misleading at worst. It failed to emphasize that Popper also wrote extensively on natural selection, Darwin's main mechanism for evolutionary change, and a mechanism clearly at issue in this case because of the court's unnecessary attempt to define science and legitimize evolution. In his writings, Popper questioned whether natural selection is a scientific theory; rather, he claimed it nothing more than an empty tautology.[10, 11, 12, 13]

This rejection of the scientific status of selection theory first became popular in creation-science circles during the late 1960s and 1970s. It started because of a common definition of natural selection as the "survival of the fittest," a phrase first used by Herbert Spencer in the middle of the nineteenth century and subsequently adopted by Darwin. Popper questioned the efficacy of this phrase. This was an important criticism of evolution as science since the New Synthesis often defined fitness in terms of survival. Popper believed that within this context, fitness means merely the survival of those that survive natural selection, thus reducing natural selection to a tautology.[10, 11, 12, 13]

For example, Popper stated[10, 11, 12, 13]: "the trouble about evolutionary theory is its tautological or almost tautological character: the difficulty is that Darwinism and natural selection, through extremely important, explain evolution by 'survival of the fittest' yet there does not seem to be much difference, if any, between the assertion 'those that survive are the fittest' and the tautology 'those that survive are those that survive.' For we have, I am afraid, no other criterion of fitness than actual survival, so that we can conclude from the fact that some organisms have survived that they were the fittest, or those best adapted..."

Not surprisingly, a number of scientists and philosophers rejected Popper's view.[10, 11, 12, 13, 14] For example, Stephen Jay Gould in a strong defense of Darwinian evolution stated that it is a response to changing environments; within those environments certain genetically determined morphological, physiological, and behavioral traits are superior as designs for survival. These traits confer fitness by an engineer's criterion of good design, according to Gould, not by the empirical fact of their survival and spread.[14]

Gould claimed that natural selection is based on the belief that statistically the fittest organisms do survive longer and reproduce more frequently; however, fitness is defined not in terms of survival, but as a measure of the organism's ability to cope with the environment by getting food, escaping predators, keeping warm, and so on. Those that have an advantage of this kind in the struggle for existence tend statistically to survive at the expense of others less favorably endowed; therefore, natural selection contributes to the spread of adaptive characters within a population.[14]

Others, Peter Bowler for example, have added nuance to this argument. He maintained that it is easy to confuse the modern concept of reproductive fitness with mere survival. In some cases, biologists cannot identify the characteristic or characteristics that are actually conferring fitness. As a result, they have often defined a particular gene as "fit" because it is maintained in the population at a high level. In these cases, Bowler acknowledged, Darwinians have fallen into a circular argument. However, he denied that this is the result of the supposed tautologous nature of natural selection. Although he conceded that the theory is sometimes impossible to test in practice.[29]

Another criticism Popper directed at natural selection had to do with its ability to be falsified. Falsifiability is an important concept in this discussion because Ruse and the court used it as a key criterion in separating science from nonscience. Popper in the late 1950s and middle 1960s argued that the scientific character of any theory is determined by its potential to be falsified by observation.[10] For example, in a 1965 paper attempting to demarcate scientific theories from nonscientific ones, he stated: "There will be well-testable theories, hardly testable theories, and non-testable theories. Those which are non-testable are of no interest to empirical scientists. They may be described as metaphysical."

According to Popper, a theory's potential for falsification can be used to assess its scientific character. A metaphysical theory is one that cannot be tested, one that is falsified. Only by fulfilling the condition of potential falsifiability is a theory testable and, therefore, scientific. It was this part of Popper's ideas that Ruse testified to during the *McLean* proceedings. However, what was not emphasized was that Popper insisted that, measured by this standard, Darwinism is untestable. At best it constitutes a "metaphysical framework" for formulating properly testable theories. This assertion is based in part on the apparently speculative nature of many Darwinian explanations of how particular structures may have evolved. Popper concluded that it is always possible to come up with some sort of adaptive explanation, although in many cases it is apparently impossible to test whether any particular explanation is valid. Further, many hypotheses are based on individual circumstances rather than general laws, which implies that Darwinism is not capable of predicting the future course of events. The lack of predictive ability again signifies that the theory does not expose itself to proper testing.[10, 11]

Popper originally put modern Darwinism in such a prescientific category. Despite its alleged scientific inadequacy, he still stressed the value of the theory by calling it a "research programme." Popper was careful, however, to state that his criterion of falsifiability is not one that separates theories according to their truthfulness. In fact, he believed that most scientific theories have their origins in untestable metaphysical theories and myths.[10]

Popper's views were subject to a number of other criticisms. For example, Imre Lakatos's analyses of scientific methodology differed from Popper's on the question of what constitutes a scientific theory. Lakatos claimed that Popper's criteria for demarcation must be altered. Lakatos developed the idea that certain types of scientific theories are protected from conflicting empirical data. He contended that scientific theories are arranged in hierarchies in which core theories are protected from refutation by the presence of auxiliary hypotheses. For example, he stated: "The basic unit of appraisal must not be an isolated theory or conjunction of theories, but rather a 'research programme' with conventionally accepted (and thus by provisional decision 'irrefutable') 'hard-core' and with a 'positive heuristic' which defines problems, outlines the construction of a belt of auxiliary hypotheses, foresees anomalies and turns them victoriously into examples, all according to a preconceived plan."[10, 15] So long as a research program is progressing, it is the "positive heuristic" that guides research rather than the anomalies, and the presence of auxiliary theories, under these circumstances, would not necessarily be interpreted as bad science within this framework.[10]

Lakatos showed that at least some scientists do not test a core theory by trying to falsify it. The complexity of structure of a hard-core theory may in the short term make simple and decisive tests (falsification) difficult, or even impossible. However, Lakatos did not deny that a rational reconstruction can be given for the growth of scientific knowledge, despite his belief that core theories are protected from direct testing—that is, from falsification.[10]

Lakatos proposed a modified version of Popper's criterion of falsifiability, suggesting rather that a theory's scientific value can be measured by its ability to produce deductive predictions, rather than its ability to be falsified. For example, he stated: "A research programme is said to be progressing as long as its theoretical growth anticipates its empirical growth, that is, as it keeps predicting novel facts with some success."[10, 15]

Lakatos believed that the Modern Synthesis of evolutionary biology also acts as a major research program in biology. The hypotheses of common ancestry, microevolution, and Darwinism are "hard-core" theories and as such are locked in an array of auxiliary hypotheses.[10, 15]

Incredibly, in his own writings before the trial, Michael Ruse himself criticized Popper. Ruse argued that the strength of Darwinism lies in its ability to link a wide range of phenomena into a comprehensive explanatory system. He asserted that the theory has made predictions about the general character of nature, which have been substantiated. Ruse conceded, however, that testable hypotheses cannot always be provided for Darwinian explanations of particular developments in the history of life. Although he pointed out that general trends in the fossil record are consistent with the theory, he also asserted that certain types of discoveries can be made that would falsify Darwinism's whole explanation of the past: for example, human fossils unambiguously found in the most ancient geological strata. Ruse claimed that although certain evolutionary laws can never be tested experimentally because of the time scale involved, such laws can theoretically be checked against future discoveries in the fossil record. He stated that even in the case of particular adaptive explanations, such types of tests are often possible.[11]

Popper, ultimately, softened his original criticism concerning the scientific status of Darwinism. In 1978 he stated: "I have in the past described the theory as 'almost tautological,' and I have tried to explain how the theory of natural selection could be untestable (as is a tautology) and yet of great scientific interest. My solution was that the doctrine of natural selection is a most useful metaphysical research programme... I still believe that natural selection works in this way as a research programme. Nevertheless, I have changed my mind about the testability and the logical status of the theory of natural selection: and I am glad to have an opportunity to make a recantation."[10, 11, 15] However, despite the recantation, many creationists co-opt Popper's original criticisms in an attempt to refute modern Darwinism.

The point of this discussion is not to give support to creationist assertions, but to emphasize that the demarcation of science from nonscience criteria, adopted dogmatically by the court, was at the time of the decision, or now, not universally accepted by scientists and philosophers. As discussed, a number of scholars have expressed considerable disagreement and concern with the *McLean* analysis. In a highly critical evaluation of Ruse's testimony, philosopher of science Larry Laudan charged Ruse with unconscionable behavior for failing to disclose the vehement disagreements among experts regarding scientific boundaries in general, and Popper's lines of demarcation in particular.[18]

Laudan believed that the *McLean* decision was the right verdict for the wrong reasons. It rests, in his opinion, on a host of misrepresentations about what science is and how it works. *McLean* contended that creationism is untestable, untentative (dogmatic), and unfalsifiable. Laudan believed that all three charges are of dubious merit. He stated that creationists make a wide range of testable assertions about empirical matters of fact. He cited, as examples, the creationist assertions that earth is of recent origin (six thousand to twenty thousand years old), that most of the features of earth's surface are the result of a Noachian (biblical) flood, and that humans and other types of organisms were created at the same time. Laudan stated that each of these examples has been tested repeatedly and, in the opinion of the vast majority of scientists who have looked at the results, have failed such tests. Laudan stated that if any doctrine in the history of science has been falsified, it is the set of claims associated with creation science.[18]

Laudan did admit, however, that some tenets of creationism are not testable in isolation; for example, the claim that humans emerged by a direct supernatural act of creation. However, he believed that this scarcely makes creationism "unscientific." It is widely acknowledged, he stated, that many scientific claims are not testable in isolation, but only when embedded in a larger system of statements, some of whose consequences can be submitted to test.[18]

Claims that scientific creationism lacks tentativeness was also criticized by Laudan. According to Laudan, the *McLean* court's claim that creationists have refused to change their views regardless of the evidence developed during the course of their investigation is mistaken. Laudan asserted that if the claims of modern-day creationists are compared with those of their nineteenth-century counterparts, significant shifts in orientation are evident. Laudan conceded that some of creationism's core assumptions, for example, that there was a Noachian flood, that humans did not evolve, or that God created the world, seem unchanged and unchangeable. However, he stated, that scientists of any generation likewise regard certain beliefs as so fundamental that they are not open to repudiation or negotiation.[18]

Laudan further took issue with the *McLean* court's opinion that science is a matter of natural law and explainable by natural law, whereas scientific creationism is not. For example, *McLean* branded as unscientific the creationist assertion that there are outer limits to the change possible within any species that cannot be explained by natural law. The court also stated that a worldwide flood, as an explanation for the earth's geology, is not a product of natural law, nor can its occurrence be explained by natural law. Laudan questioned sarcastically how the *McLean* Judge (Judge Overton) knew these things![18]

The real issue to Laudan was not whether creationism satisfies some undemanding and highly controversial definition of what science is. The real question, according to Laudan, is whether the existing evidence provides stronger arguments for evolutionary theory than creationism. Laudan dismissed the problem of demarking science from nonscience as a "red herring," and he labeled it a "pseudoproblem." To him, critics of creationism do not help themselves by pretending that science is characterized by an uncompromising open-mindedness[18] that they know does not exist.

Ruse took issue with Laudan's views, and he issued a strong and pragmatic rebuttal. He asserted that the issue of what constitutes science arose because creationists claim their ideas qualify as genuine science, rather than as Fundamentalist religion. Ruse claimed that certain attorneys developing the case believed it important to show that creation science is not genuine science. However, according to Laudan, creation science is weak science and, as such, should not be taught. Ruse maintained that the federal constitution does not bar the teaching of weak science; it only bars, through the Establishment Clause, the teaching of religion. Therefore, the plaintiffs' tactic in the *McLean* case was to show that creation science is less than weak science; it is not science at all.[19] What Ruse's arguments avoids, however, is that intellectual integrity is just as important for those who oppose religion being inserted into the public-school science curriculum as it is for those who dishonestly attempt to insert it—in fact, maybe more so![20] We have, through history, come to expect duplicity and intellectual dishonesty from creationists in regard to the evolution in public-school cases. Have we also learned the same about those opposing them?

This issue is a major problem in a whole line of evolution in public-school cases, from *Scopes*, to *Epperson*, to *McLean;* and as will be seen in subsequent chapters, in *Edwards v. Aguillard* and in *Kitzmiller v. Dover Area School District*.[21, 22] That is the less-than-honest arguments of participants on both sides designed to support their particular position. One wonders if these cases act as a model to better understand a pervasive problem associated with the legal and judicial systems in this country?

Ruse attempted to minimize the demarcation problem. Although Ruse admitted that his criteria for demarcating science from nonscience are often not sufficient[19, 23] and that the five criteria listed in the *McLean* opinion do not always demarcate adequately, he maintained that they often do. Ruse contended that the slightest acquaintance with the creation science literature and the creationist movement show that creation science fails as science. For example, Ruse cited a passage by a leading creationist, Duane Gish, in *Evolution: The Fossils Say No!*: "CREATION. By creation we mean the bringing into being by a supernatural creator of the basic kinds of plants and animals by the process of sudden, or fiat, creation. We do not know how the Creator, created, what processes He used, for He used processes which are not now operating anywhere in the natural universe… We cannot discover by scientific investigation anything about the creative processes used by the Creator."

Ruse further cited a passage by Henry Morris, who Ruse labels as the founder of the creation-science movement: "it is…quite impossible to determine anything about Creation through a study of present processes, because present processes are not created in character. If man wishes to know anything about Creation (the time of Creation, the duration of Creation, the order of Creation, the methods of Creation, or anything else) his sole source of information is that of divine revelation…we are completely limited to what God has seen fit to tell us…."

Ruse, therefore, stated that creation scientists, by their own words, "admit that they appeal to phenomenon not covered or explicable by any laws that humans can grasp as laws. It is not simply that the pertinent laws are not known."[19]

Ruse also took issue with any assertion that creation science is tentative or testable. For example, he cited the contents of a document that prospective members of a chief creation science organization, the Creation Research Society, must sign specifying their beliefs: "The Bible is the written Word of God, and because we believe it to be inspired throughout, all of its assertions are historically and scientifically true in all the original autographs. To the student of nature, this means that the account of origins in *Genesis* is a factual presentation of simple historical truths… Finally, we are an organization of Christian men of science, who accept Jesus Christ as our Lord and Savior. The account of the special creation of Adam and Eve as one man and one woman, and their subsequent fall into sin, is the basis for our belief in the necessity of a Savior for all mankind. Therefore, salvation can come only thru accepting Jesus Christ as our Savior."[19]

Finally, Ruse challenged Laudan's claim that some parts of creation science are falsifiable and other parts are revisable. Ruse asserted that such parts are not falsifiable or revisable in a way indicative of genuine science. For example, he contended that creation science exists solely in the imaginations and writings of a relatively small group of people. He further contended that their publications show that there is no way they will relinquish belief in the flood, whatever the evidence. In this sense, their doctrines, according to Ruse, are unfalsifiable.[19]

Creationists obviously took issue with these assertions. They contended that they produced an alternative model for origins and how organisms got here. In support of their methodology, they relied in part on philosopher Thomas Kuhn. Kuhn, in his influential book *The Structure of Scientific Revolutions* in 1962, described scientific development partially in terms of completing models or paradigms (a common conceptual framework), as he called them, rather than the accumulation of objective knowledge. Differing in certain respects from Popper's views on scientific methodology, Kuhn believed that the central theory in a research program is protected, by various mechanisms, from being tested critically. Kuhn claimed that normal scientific research proceeds within the bounds of a paradigm, which is shared by the workers involved in a particular field. The specific paradigm held will dictate to the scientists which questions are worth asking and what problems need be solved. Kuhn maintained that a paradigm is not displaced in the normal course of science because it is falsified only when mounting anomalies within the theory, combined with the appearance of novel discoveries, lead to the production of rival paradigms. Within the course of normal scientific activity, the criterion for falsifiability of individual predictions is considered appropriate, and the outcome of such tests directly determines the fate of the hypothesis concerned. However, according to Kuhn, the larger paradigm, which guides the course of normal scientific research, cannot be tested in this way. Indeed, Kuhn maintained that a scientist's choice between rival paradigms is initially made as an act of faith. In Kuhn's scheme, scientific revolutionaries, even while constituting a small minority, occupy scientific ground.[17, 24] Therefore, creationists saw no reason why their model of origins should not be accorded space within the scientific curriculum.[17]

Ronald Numbers also took issue with ideas of the type espoused by Ruse. Numbers, relating the opinion of certain sociologists of science, contended that the *McLean* trial provides a revealing glimpse of scientists vigilantly guarding their boundaries. Numbers believed that just as creationists pushed the limits of science to accommodate their religiously inspired agenda, so their opponents invoked a narrow definition of science to maintain their "monopoly over the market for "scientific" knowledge in Arkansas schoolrooms. Numbers suggested that by discrediting the creationists as "pseudoscientists" unworthy of public patronage, the established scientific community hoped to eliminate a politically powerful competitor for scarce resources. Numbers stated that whatever the respective merits of the two sides, their struggle illustrates the historically contingent nature of "science" and the futility of assigning to the term an invariant meaning.[17] This statement is supported by the changing nature of natural science over time. Before and during the nineteenth century, and even into the early part of the twentieth century, many natural scientists did not exclude supernatural explanations for certain biological phenomena.

The *McLean* trial, according to Numbers, also shattered myths about the so-called warfare between science and religion. As the *McLean* court pointed out, the plaintiffs were a scientific society, a teacher, a science teachers' organization, various clergy from a variety of Christian denominations, and various Jewish organizations. No religious group appeared on the list of defendants. Further, testifying at the trial against creationism were Christian clergypersons. In contrast, most of the witnesses in support of the act were well-credentialed scientists. Numbers related that, given the composition of the two sides, Langdon Gilkey, a theologian and participant, characterized the controversy as involving "two bizarre, unaccustomed and visibly uneasy partnerships: on the one side a union of what we might call elite religion and elite science, and on the other side a union of 'popular' (fundamentalist) religion with 'popular science.'"[17s]

Entering into this controversial area of philosophy of science was not the only audacious part of the *McLean* opinion. Incredibly, the court even chose to decide what topics should be considered under the rubric of evolutionary theory. Most particularly, the court stated that origin-of-life topics do not qualify under the domain of evolution.[1] Although this position by the court may be held by some, it also is not universally accepted.

On a technical level, two broad areas of investigation emerged in the twentieth century, one interested in the mechanisms and basis of evolutionary change, a second interested in origin of life from nonlife questions and mechanisms. In many respects, these two areas have a different history, and deal with somewhat different issues. Nonetheless, excluding origin-of-life questions from evolutionary theory is the result of another oversimplified analysis by the *McLean* court. Although there is some support for this position in the present scientific community,[1] most other scientists and historians of science think otherwise. Biologists in the latter part of the eighteenth century and early part of the nineteenth did not make scientific origin of life topics a central issue in evolutionary theory.[25] In *Origin*, Darwin wrote as though the first forms of life may have been divinely created as a way of evading the issue. The concept of spontaneous generation of life from nonlife was not popular in the early nineteenth century, and Darwin did not want his theory tied to any wild speculations about totally unknown causes. However, some of Darwin's supporters realized that to be consistent, and that evolutionary theory ought to explain in terms of natural causes the origin as well as the subsequent development of life. Both Thomas Huxley and Ernst Haeckel set forth theories linking living and nonliving matter.[11, 26, 27] Darwin himself stated in a letter unpublished during his lifetime that "It is often said that all the conditions for the first production of a living organism are now present, which could ever have been present. But if (and oh! what a big if!) we could conceive in some warm little pond, with all sorts of ammonia and phosphoric salts, lights, heat, electricity, etc. present, that a protein compound was chemically formed ready to undergo still more complex changes, at the present day such matter would be instantly devoured or absorbed, which would not have been the case before living creatures were formed."[11, 26, 27]

Peter Bowler stated that the appearance of the first modern theory of the origin of life on earth actually helped boost the revival of Darwinism during the New Synthesis. The ancient belief in the spontaneous generation of life was finally completely discredited in the late nineteenth century by Louis Pasteur's famous swan-neck flask experiment in the 1880s that showed there is no evidence for life being formed from nonliving matter. This breakthrough, however, left science with no plausible naturalistic explanation of how the life first began. In 1936, the Russian biologist Alexander Oparin introduced a new approach to the question of the origin of life. Oparin ultimately developed an idea that replaced the concept of spontaneous generation with a concept of chemical evolution through levels of increasingly complex organization. According to Bowler, this not only extended the range of evolutionary thinking, but also gave renewed support to Darwinism, because Oparin postulated a form of natural selection as the mechanism of gradual improvement among preliving structures.[11, 26, 27]

Oparin's theory was presented in his book, *The Origin of Life*, published in 1936, and translated into English in 1938. He assumed that early earth was originally a sterile planet with no free oxygen in the atmosphere. If free oxygen had been present, it would have oxidized the chemicals necessary for the appearance of life, making them unable to participate in life's emergence. Instead, Oparin believed that the early atmosphere of earth was a reducing environment, containing hydrocarbons and ammonia with virtually no free oxygen. Chemical reactions among these gases produced complex organic molecules that dissolved in the oceans and survived in a manner that allowed them to form a rich "primordial soup." The organic molecules continued to combine to form polymolecular open system or coacervates (minute droplets with a definite structure) capable of absorbing materials from their surroundings. Natural selection came into play within Oparin's scheme. Those coacervates with more stable structure survived at the expense of the less stable. To Oparin, the more successful structures became the "prebionts," the intermediate stages before the appearance of the true life in the form of the first cells. Oparin at first saw the origin of life as a unique event, through the accidental combination of chemicals, occurring when the first cells were precipitated from a colloidal suspension. However, he ultimately moved away from the idea of a single-step process for the formation of life based on chance combinations. Instead, he adopted a far more sophisticated view with no sharp dividing line marking the appearance of life. To Oparin, the process of chemical evolution gradually increased the level of organization, providing a complete continuity between inorganic matter and the first living cells as part of a process of selection. Oparin argued that once life appeared on the earth, the process could never be repeated, since living organisms themselves would destroy the earlier stages in the process as soon as they occurred. Therefore, it was impossible for life to be created now, just as Pasteur had shown.[11, 26, 27]

In 1953, a young scientist, Stanley Miller, performed experiments that partially confirmed certain early steps of Oparin's theory. Miller passed an electric current (to mimic an early earth energy source, lightning) through a sample of the reducing atmosphere (lack of free oxygen) suggested by Oparin, with extremely interesting results. Miller's experiment yielded the synthesis of certain amino acids, the building blocks of proteins, providing a type of evidence in support of Oparin's views.[11, 26, 27]

Many laboratories since 1953 did and do Miller-type experiments, using a variety of recipes for the proposed reducing, oxygen-free, atmosphere of early earth, and using a variety of energy sources. Abiotic synthesis of organic compounds has been a common feature of these experiments. Laboratory analogs of the primeval earth have been used to make all twenty naturally occurring amino acids commonly found in organisms, several sugars, lipids, certain components of DNA and RNA, and other molecules important to living cells.[28]

Bowler and Farley contended that when Miller and others provided experimental support for some early steps of the Oparin hypothesis, Oparin's views of the origin of life became accepted as an integral component of modern evolutionism.[11] Although in the latter part of the twentieth century and early twenty-first century origin-of-life topics have generated considerable scientific controversy, particularly in the details of chemical evolution and whether proteins, or one or another type of nucleic acid, such as DNA or RNA played the initial role. Controversy also exists as to whether the absence of free oxygen in the early earth atmosphere was necessary for life to naturally emerge and were the presence of clays necessary. Such topics have generally been included in evolutionary textbooks[11] and general biology textbooks dealing with evolution. To a large degree, origin-of-life topics is presented as a rational extension of a naturalistic evolutionary philosophy and the mechanism of natural selection.

In addition to the controversial, political, and ideological rulings on the definition of science and the appropriate contents of evolutionary theory, the *McLean* court also entered the difficult scientific terrain of deciding what mechanisms actually drive evolutionary change, stating that natural selection and mutation are insufficient. Although this position reflects the current understanding of the majority of evolutionary biologists and the understanding at the time of *McLean,* and now, one must question whether it is prudent for any federal court to enter such a technical and specialized area without the obvious expertise to do. It appears that the *McLean* court accepted without question the scientific testimony presented by plaintiffs' witnesses, while disregarding the testimony of creationists and their witnesses. In this regard, the bias inherent in the *McLean* decision was strikingly similar to those of *Epperson* and *Wright.*

None of these controversies needed to be created. Since the *McLean* court rightly concluded that the Balanced Treatment Act's purpose was at minimum primarily religious and it would entangle school boards and courts to an unmanageable extent in monitoring books and curricula in multiple topics that contain evolutionary ideas, the court did not need to tread into the dangerous intellectual and legal waters of what constitutes science or what topics are legitimately parts of evolution theory. The court should have confined the limits of its decision as narrowly as possible. The court should have crafted a legally narrow, judicially constrained decision instead of one that is obviously a thinly veiled, ideologically laden attempt with questionable integrity, to discredit scientific creationism.

The *McLean* decision's obvious intent was to establish the scientific legitimacy of evolutionary theory, while demonstrating that creation science fails to meet any reasonable criteria for inclusion as a scientific theory. In this effort, the *McLean* court strayed far beyond any previous federal court decision or, for that matter, what was prudent. Creation science was held in *McLean* to be merely a pretext for the propagation of Fundamentalist religious beliefs and the Arkansas Statute was held to be unconstitutional as violating the Establishment Clause of the First Amendment.

The *McLean* decision unambiguously established that balanced treatment statutes faced a tough Establishment Clause challenge in the Federal courts. Notwithstanding this, balanced treatment legislation and litigation did not end. In Louisiana, a related Statute was subsequently enacted by the State legislature and was quickly challenged in the Federal courts, ultimately in the Supreme Court of the United States, in *Edwards v. Aguillard.*[21]

FEDERAL EVOLUTION LITIGATION IN THE 1980S: A LOUISIANA BALANCED TREATMENT STATUTE

Once Again, the Supreme Court Addresses Evolution
Edwards v. Aguillard

In 1987, five years after the *McLean* decision in an Arkansas Federal District Court, a similar balanced-treatment act, one from Louisiana, was decided by the Supreme Court of the United States in *Edwards v. Aguillard.*[1] This was the second time in American judicial history that a case involving evolution in the public schools was litigated in the Supreme Court, *Epperson v. Arkansas* being the first. The Arkansas Act litigated in *McLean* and the Louisiana Act litigated in *Edwards* came from the same model that was prepared around 1977 by Paul Ellswanger, and circulated among various individuals throughout the country.[1, 2] However, in one significant respect the two acts were different. The act relevant to *Edwards* did not include detailed definitions of creation science and evolution as did the Arkansas Act. Consequently, a court analyzing the definition of creation science, as defined in the Louisiana *Edwards* Act, would not be able to equate it directly with the biblical account of creation, as was done in *McLean.*[3]

The *Edwards* case addressed the constitutionality of an act officially titled the Louisiana Balanced Treatment for Creation-Science and Evolution-Science in Public School Instruction Act.[3] The Creationism Act, as it is often called, was enacted in 1981. As a balanced treatment statute, it forbade the teaching of evolution in public schools unless creation science was also taught. No school was required to teach evolution or creation science; however, if one was taught, the other also had to be taught.

The Creationism Act defined creation science merely as "the scientific evidence for creation and inferences from those scientific evidences" and defined evolution science as "the scientific evidence for evolution and inferences from those scientific evidences." No other definitions were given for either. As stated, this was a significant departure from the statute litigated in *McLean*, which provided detailed definitions of both, with the definition of creation science contributing significantly to the *McLean* statute's failure in court.

As in *McLean*, legal action in *Edwards* was initiated by an eclectic group of Louisiana parents, teachers, and mainstream religious leaders, challenging the act's constitutionality in a federal district court.[1] The Louisiana officials responsible for implementing the act were defendants. One thing different procedurally about this case compared with *McLean* was that an evidentiary trial was never held. In *McLean*, there was a full, lengthy, and complicated evidentiary trial with multiple witnesses, and the court's decision was never appealed. Instead in *Edwards*, a summary judgment was granted by the district court without an evidentiary trial and that decision was appealed all the way to the Supreme Court of the United States. There was never an evidentiary trial held in *Edwards*.

Again, the *Edwards* district court granted a summary judgment to the plaintiffs[1]; that is, a final court decision without a trial. Rule 56 of the Federal Rules of Civil Procedure provides for summary judgments. According to Rule 56, in order for a party to be granted a summary judgment, that party must show that there is no genuine dispute as to any material fact, and that the party is entitled to judgment as a matter of law. This is an extraordinary remedy because a court renders final decision against one of the parties without an evidentiary trial.

The basis for the *Edwards* decision is a familiar one for cases involving the teaching of evolution in public schools; since *Epperson*, they had been decided on First Amendment grounds. The district court held the act unconstitutional in violation of the Establishment

Clause of the First Amendment to the federal constitution[1] as a matter of law without the need for a trial.

The *Edwards* district court decision was appealed to the Court of Appeals, which affirmed the summary judgment. The Court of Appeals held that the act's stated purpose of protecting academic freedom was inconsistent with the act's requirement that creation science be taught whenever evolution was taught. The Appeals Court stated that the plain language of the statute made its purpose predominantly religious and the lack of secular purpose apparent. The court felt that academic freedom, as that word is universally defined, allows individual instructors, in the exercise of their legitimate academic functions, to teach what they consider appropriate about a subject. The court concluded that the act was contrary to the very concept it says it supports by requiring the teaching of creation science whenever evolution is taught.[4]

As an aside, it's ironic hypocritical that in 1994, a Ninth District Federal Court of Appeals did limit the right of teachers to teach "what they consider appropriate about a subject" by upholding, in *Peloza v. Capistrano School District*, a district court's decision that a public school can constitutionally require a science teacher to teach evolution in biology class. The court, in effect, ruled that evolution is science and it held that a teacher does not have any constitutional right or academic freedom right to teach what he or she believed appropriate about the question of origins of species even if the teacher's belief is that evolution is not appropriate or scientifically justified.[25] Academic freedom as it is "universally defined" obviously didn't have quite the significance or importance to a federal court in this situation.

Back to *Edwards*. The Appeals Court held that the purpose and intent of the Louisiana legislature in enacting the act was to discredit evolution by counterbalancing its teaching at every turn with the teaching of creationism, which the court, as did *Epperson* and *McLean* before it, considered a religion. Applying the *Lemon* test's purpose prong, the Appeals Court concluded that the act violated the Establishment Clause of the federal constitution. The court stressed that its decision was not made in a vacuum, stating that historically supporters of the teaching of creationism have strong Fundamentalist religious links.[4]

A panel of three Appeals Court judges initially heard and decided the case. A petition for rehearing *en banc* was filed requesting reconsideration of the decision in an effort to overturn it. The petition was considered by the entire Appeals Court consisting of fifteen judges. The request for rehearing was denied by a close eight-to-seven vote. Therefore, the summary-judgment ruling of the three-judge panel was upheld.

A sharp dissenting opinion was filed by Judge Gee and joined by six additional Appeals Court judges. In a written dissent to the Appeals Court ruling, Gee stated that the act did not violate the Establishment Clause of the federal constitution. He stated that the legislative record contained affidavits from highly qualified scientists affirming that evolution is not a fact and that there is evidence that life and the universe came about in a different manner.[5] Gee believed that this alone showed a genuine dispute of material facts in the case, mandating that a summary judgment not be entered.

Gee also rejected the court's ruling that the act violated the purpose prong of the *Lemon* test. He stated that the act articulated a secular, rather than a religious, purpose; that is, to protect academic freedom. Therefore, he believed that the court should not have invalidated the act for violating the First Amendment. He believed that the Appeals Court disregarded the statute's clear words and the legislative statements of secular purpose, and it made improper assumptions regarding the motives of the legislators.[5]

An appeal was taken to the Supreme Court of the United States by defendants.[1] The Supreme Court framed the issue narrowly as whether the act violated the Establishment Clause within the special context of the public-school system.[1] In a contentious decision, a majority of seven out of nine Supreme Court judges affirmed the Appeals Court decision to grant summary judgment holding that mandating balanced treatment of creation science with evolution in the public-school science curriculum is an unconstitutional action on the part of Louisiana. Nonetheless, Justice Scalia, joined by Justice

Rehnquist, wrote a scathing dissent along similar lines as the Appeals Court dissent by Judge Gee.

The reason the Supreme Court framed the issue as it did was because the court historically has given public schools considerable discretion in developing curricula as they see fit. Yet despite this, the court has been vigilant in monitoring compliance with the Establishment Clause, often invalidating various types of statutes that advance religion in public schools. The court articulated that the reason for the vigilance is because students of the relevant age group are very impressionable and that school attendance is not voluntary for them.[1]

A Supreme Court majority of seven of nine judges applied the first prong of the *Lemon* test in its analysis of the Creationism Act, the so-called purpose prong.[1,6] In applying *Lemon,* the majority articulated its standard of review. Citing Justice O'Connor's concurring opinion in *Lynch v. Donnelly*, the court wrote that a proper standard of review is whether the actual purpose of a statute is to "endorse" or disapprove religion.[1,7] This standard would in subsequent Supreme Court decisions develop into what is now called the endorsement test for First Amendment constitutionality. This test will be important in the final case discussed in this work, a case looked at and analyzed in the next chapter, *Kitzmiller v. Dover Area School District*.[8] The majority also cited standards articulated in *Wallace v. Jaffee*.[9] In *Wallace*, the majority held that government's intention to promote religion is demonstrated when a state enacts a law to serve a religious purpose, either to promote religion in general or, as articulated in *Stone v. Graham*[10], to advance a particular religion. The majority asserted that if it concluded that a law was enacted with the purpose of endorsing religion, no further consideration of the second or third prong of the *Lemon* test is necessary to invalidate that law under the Establishment Clause.[1]

The majority stated that although the Supreme Court is deferential to a state's articulation of secular purpose, that statement of purpose must be sincere and not a sham. The majority felt that the Creationism Act's stated purpose of protecting academic freedom lacked requisite sincerity. The act defined "protecting academic freedom" in a very unusual way, teaching all the evidence with respect to the origin of human beings. Almost universally, academic freedom is defined as a right by teachers to teach what they choose concerning their particular subject. Therefore, the act's true purpose seemed to diminish academic freedom by removing the flexibility teachers had to teach evolution without also teaching creation science. It also decreased the flexibility of teachers to supplement the curriculum with alternative theories to evolution besides creationism.[1]

The Supreme Court majority did not believe that the legislative history of the act supported a purpose of protecting academic freedom. After reviewing the act's legislative history, the majority concluded that legislative sponsor, Senator Bill Keith's, real purpose was to narrow the science curriculum and to discredit evolution. In support, the majority asserted that during the legislative hearings, Senator Keith stated that his preference would be that neither evolution nor creationism be taught in the public schools.[1]

Further, the majority stated that the act violated its goal of basic fairness by showing a discriminatory preference for creation science over evolution. The majority supported this by stating that the act required development of curriculum guides in creation science but not in evolution and required the formation of resource services for teachers in creation science but not in evolution. In addition, only creation scientists could be on the resource panels. The act also prohibited school boards from discriminating against creation scientists but failed to protect those who teach evolution.[1] The Supreme Court majority believed that these factors demonstrated that the historic and contemporaneous antagonisms shown by Fundamentalists toward evolution were present in this case as much as they were present in *Scopes, Epperson*, and other like cases. The majority held that the preeminent purpose of the Louisiana legislature in enacting the act was to advance religion.[1]

The term "creation science" was viewed by the Supreme Court majority as a Fundamentalist religious doctrine. The majority stated that even the leading expert witness supporting creation science, Edward Boudreaux, testified at the legislative hearing on the act that creation science included a belief in the existence of a supernatural creator. Act sponsor Senator Keith explained during the legislative hearings that his opposition to evolution was the result of his religious beliefs. Keith stated that since evolution was harmonious with the principles of secular humanism, theological liberalism, religious humanism, and other philosophies and religions, the scientific evidence supporting his religious beliefs should also be included in the public-school curriculum. The majority noted that Keith also testified that most creation scientists in the United States were affiliated with one of two organizations, the Institute for Creation Research or the Creation Research Society. Both organizations, the majority concluded, were religious in purpose. To support this contention, the majority reviewed statements and membership requirements from both organizations. The Institute for Creation Research claimed to be established to address the need for the United States to return to a belief in a personal, omnipotent creator, who has a purpose for His creation and to whom all people must give account. A goal of the institute is a revival of belief in special creation as the true explanation of the origin of the world. The Creation Research Society, the majority continued, requires members to subscribe to the belief that the Bible is the written word of God and that all of its assertions are historically and scientifically true. In order to study creation at the Creation Research Society, a member must accept a literal interpretation of *Genesis*.[1]

The Supreme Court majority concluded that the purpose of the act was to restructure the science curriculum to a particular religious viewpoint. The majority stated that out of many possible science subjects taught in the public schools, the Arkansas legislature chose to affect the teaching of the one scientific theory that historically has been opposed by certain religious sects. In the majority's view, the act was designed either to promote the theory of creation science, which embodies a particular religious tenet, by requiring creation science to be taught whenever evolution is taught, or by prohibiting the teaching of a scientific theory, disfavored by certain religious sects. The majority held that either activity is a violation of the Establishment Clause.[1]

The majority did acknowledge that scientific critiques of the prevailing theories of origins might be taught in public schools if it were done with the clear secular intent of enhancing the effectiveness of scientific education. But the majority stated that because the primary purpose of the Creationism Act was to endorse a particular religious doctrine, the act furthered religion in violation of the Establishment Clause.[1]

Justice Powell, joined by Justice O'Connor, wrote a concurring opinion. Powell reaffirmed that state and local public schools are afforded broad discretion in developing curricula. However, a statute effecting curriculum must have a valid secular purpose to pass constitutional muster. This is determined by applying the *Lemon* test. In doing so, Powell concluded that the purpose of the act was to advance religion. He believed that the statute's requirement that public schools must present scientific evidence to support creationism whenever they present the scientific evidence to support evolution is manifestly religious.[1]

In deciding this issue, Powell used a traditional approach to judicial statutory construction; that is, unless otherwise defined, words will be interpreted using ordinary, contemporary, common meaning. Using this criterion, Powell stated that the theory of creation is defined by *Webster's Third New International Dictionary* (unabridged 1981) as holding that matter, the various forms of life, and the world were created by a transcendent God out of nothing. Powell believed that concepts concerning God or a Supreme Being of some sort are manifestly religious and do not become less religious because they are presented as a philosophy or science.[1]

However, Powell conceded that a religious purpose is not enough to invalidate an act of a state legislature; the religious purpose must predominate, in his opinion. Powell maintained that the stated purpose of the act at issue in *Edwards*, to protect academic freedom, rendered the purpose at least ambiguous. Therefore, he turned to the legislative history of the act for guidance, concluding tshat the model act upon which the Louisiana statute relied was also the basis of the Arkansas Act litigated in *McLean*. Citing the *McLean* decision, Powell stated that the Louisiana Act in both its concepts and wording convey inescapable religiosity. Powell concluded that the Louisiana legislature acted with the unconstitutional purpose of structuring the public-school curriculum to make it compatible with a particular religious belief. Therefore, he reiterated that the purpose of the act was to advance a particular religious belief.[1]

Justice White also wrote a concurring opinion. Showing deference to the judgments of the district court and Court of Appeals, and considering the act's plain language, White concluded that the Louisiana legislature's primary purpose was to advance religion and therefore was unconstitutional under *Lemon*.[1]

It is clear from this discussion, that the Supreme Court majority decided this case based upon its conclusion that the Creationism Act was enacted by the Louisiana legislature with a clear and predominant religious purpose. The court ratified that a predominant religious purpose was sufficient to deem a law unconstitutional as violative of the First Amendment.

The Supreme Court majority opinion was not without dissent. A scathing dissenting opinion was written by Justice Scalia and joined by Chief Justice Rehnquist.[1] Scalia used technical legal arguments to justify his position. He stated that there is no legal justification for the Supreme Court's decision to affirm the Appeals Court's summary judgment.[1] Scalia stated that a summary judgment should not have been granted in this case. He maintained that a multitude of factual issues remained, incapable of resolution without a full evidentiary trial. For example, he stated that the Louisiana Supreme Court was never given an opportunity to interpret the meaning of the act before the summary judgment was granted, nor had state officials attempted to implement the act, and there was never an evidentiary hearing on the act in a state court or lower federal court before the decision.

Scalia observed that the parties to the litigation were sharply divided over the definition of creation science. One side asserted that it is a collection of educationally valuable scientific data that supports the idea that the physical universe and life appeared suddenly, and they have not changed substantially since appearing. The other side contended that creation science is not science but thinly veiled religion. Both interpretations, Scalia believed, were supported by certain aspects of the legislative history and both interpretations had several experts that testified in support. Scalia added that at least five experts called by supporters of the act testified that creation science was strictly a scientific concept that could be presented without religious reference.[1] Based upon this, Scalia argued that a summary judgment was improperly entered and that an evidentiary trial should be required before any court decision on this matter.

In addition, Scalia criticized the majority's use of the first prong of the *Lemon* test, the purpose prong. He believed that it was applied incorrectly in this case, questioning whether a law can be invalidated under the Establishment Clause on the basis of motivation alone, without regard to its effects. This broke with how federal courts have been deciding evolution and public-school cases since *Epperson* and harkened back to *Scopes*.

Scalia asserted that only once before deciding *Lemon*, and twice since *Lemon*, had the Supreme Court invalidated a law for lack of secular purpose, while listing a long line of purpose cases where state actions were held constitutional.[6] In reviewing numerous Supreme Court cases where the purpose requirement of the *Lemon* test was applied,[1, 6, 9, 10, 11] Scalia observed that some guiding principles emerged. First, regardless of what legislative purpose may mean in other contexts, for the *Lemon* test, it means the actual motives of those responsible for the challenged action. Therefore, Scalia observed, if those legislators who supported the act had a sincere secular purpose, the act survives the first prong of the *Lemon* test. Second, he stated that invalidation under the purpose test was appropriate only when there was no question that the statute was motivated wholly by religious considerations. This was an issue ignored in *McLean;* that is, did invalidation under the purpose prong of *Lemon* require that the religious purpose be the sole purpose, or was there a lesser standard possible to invalidate a law?

In all three cases cited by Scalia in which the Supreme Court struck down laws under the Establishment Clause for lack of secular purpose, the court found that the legislature's sole motive was to promote religion. Scalia, therefore, concluded that the invalidation of the Balanced Treatment Act was defensible in *Edwards* only if the record indicated that the Louisiana legislature had no secular purpose whatsoever.[1] As stated, this was a key question ignored by the *McLean* court.

While Scalia believed that the legislative purpose forbidden by *Lemon* was to advance religion, he did not believe that it forbade an individual legislator from acting on his or her religious convictions. For example, he asserted that the court would not strike down a law providing money for the hungry or shelter for the homeless if it could be demonstrated that, but for the religious beliefs of the legislators, the funds would not have been approved. He also stated that political activism by the religiously motivated individuals is part of our national heritage. He asserted that the sole purpose of a law should not be presumed to advance religion just because the law is strongly supported by organized religions or a particular faith. To do so, he concluded, would deprive religious people of their right to participate in the political process. Further, Scalia asserted that the majority should not presume that a law's purpose is to advance religion merely because it happens to coincide or harmonize with the tenets of some or all religions or to even benefit religion. Therefore, by extension, Scalia concluded that the fact that creation science coincides with the beliefs of certain religions, a fact upon which the majority heavily relied in its decision to affirm the summary judgment, should not in itself justify striking the act.[1]

Scalia further suggested that the federal constitution does not always prohibit the government from acting to advance religion. He cited at least two circumstances where government must act to advance religion, and a third where it may do so. First, if a state discovered that its employees are inhibiting religion, the state must take steps to prevent them from doing so, even though the purpose would be to advance religion, because the Establishment Clause prohibits not only a state acting to advance religion, but also a state action that is hostile toward religion. Therefore, if the Louisiana legislature sincerely believed that the state's science teachers were being hostile to religion, Scalia contended that the state could constitutionally act to eliminate that hostility without running afoul of *Lemon's* purpose test. Second, Scalia observed that sometimes intentional governmental advancement of religion is required by the Free Exercise Clause of the First Amendment[1], the clause prohibiting government interference with the free exercise of religion.

Scalia believed that the Supreme Court had not adequately reconciled *Lemon* and the court's free-exercise cases. He stated that if members of the Louisiana legislature believed that approval of the act was required by the Free Exercise Clause, it would not be unconstitutional for the legislature to enact the act, possibly even if its sole motive was to advance religion.[1]

Scalia contended that the majority had little information upon which to judge the motives of those who supported the act. For him, the only relevant evidence available was the statute itself and transcripts of seven legislative-committee hearings. Based on an analysis of this evidence, Scalia concluded that the majority was wrong in holding that the Creationism Act was without secular purpose. Scalia stated that after two hearing by the State Senate Committee on Education, the bill's sponsor, State Senator Keith, asked that the bill be referred to a study commission composed of members of both houses of the state legislature. The joint commission met twice and heard testimony. The sponsor introduced it again when the legislature reconvened. The State Senate Committee on Education met twice more and approved the bill after substantially amending it (in part over the sponsor's objection). After approval by the full State Senate, the bill was sent to the House Committee on Education. That committee conducted a lengthy hearing, adopted further amendments, and sent the bill to the full House where it was passed. The Senate then agreed on the House amendments, and the bill was signed by the governor into law in July, 1981 more than a year after it was initially introduced. Scalia related that during the bill's consideration, the sponsor did not always approve of the decisions being made by the legislature concerning the bill. Further, the legislators were aware of the bill's potential constitutional problems.[1]

Scalia also contended that his views and the views of the court concerning evolution and creation science should be irrelevant to their decision. What is important, he stated, were the views of the members of the Louisiana legislature. The vast majority of them voted to approve the bill, which contained a stated explicit secular purpose (protection of academic freedom). What was crucial for constitutional purposes, Scalia asserted, was not their wisdom, but their sincerity in believing it. Scalia related that most testimony came from the sponsor himself and from scientists and educators he presented to the legislature, many of whom had impressive academic credentials. Scalia observed that the sponsor and his witnesses testified that there are only two scientific explanations for the origin of life, evolution and creation science. Scalia contended that both theories are subject to empirical testing, and both are bona fide sciences.[1]

Scalia stated that the sponsor of the act and the witnesses contended that evolution hypothesizes that life arose out of inanimate chemical compounds and has gradually evolved over millions of years, whereas creation science hypothesizes that all life-forms now on earth appeared suddenly and relatively recently and have changed little. Since creation scientists assert that there are only two possible explanations for origins of life, any evidence against evolution, in their view, by necessity tends to prove the theory of creation science and vice versa. The sponsor and witnesses contended that the body of evidence supporting creation science is at least as strong as that supporting evolution. They believe that evolution cannot be observed in a laboratory, that the evidence for it is not compelling, and that evolution is not a scientific fact.[1] What is incredible about Scalia's view on this matter is that it ignores the dozens of origin myths that are part of many different belief systems from cultures around the world. Common knowledge and common sense argues against a two-model view of origins.

Scalia continued that the sponsor and his witnesses contended that creation science has educational value; students exposed to it develop a greater understanding of the current state of scientific evidence about the origin of life, including the evidence for evolution. They contended that creation science can and should be presented to children without any religious content. They stated that although creation science is educationally valuable and strictly scientific, it is currently being censored or misrepresented in the public schools, while evolution is held out as an absolute truth.[1] In addition, the sponsor and the witnesses contended that the censorship of creation science has at least two harmful effects: it deprives students of knowledge of one of two scientific explanations for the origin of life, and leads it them to believe that evolution is a proven fact. In effect, students are taught that science has proven their religious beliefs false. The sponsor and witnesses contended that belief in evolution is a central tenet of secular humanism, a philosophy held to be a religion by the Supreme Court. To them, public-school teachers, by censoring creation science and teaching evolution, are advancing religion in violation of the Establishment Clause.[1]

Scalia believed that the majority misrepresented the legislative history of the act. He pointed out that the legislative sponsor of the act repeatedly and vehemently denied that his purpose was to advance a particular religious doctrine, and many witnesses testified that creation science was educationally valuable and could be taught without reference to religion. Further, Scalia observed that the act had a stated secular purpose of protecting academic freedom, which was defined as students' freedom from indoctrination. The Louisiana legislature, in Scalia's opinion, did not care whether the topic of origins was taught; it simply wished to ensure that when the topic was taught, students would receive all of the evidence. Scalia questioned how the majority could legitimately conclude that this purpose represented a desire to restructure the science curriculum to conform to a particular religion. Scalia stated that the majority had no basis on record to conclude that creation science, as defined by the act, is anything other than a collection of scientific data supporting the theory that life abruptly appeared on earth, without reference to a creator. However, he stated that even if this were not so, to posit a past creator is not to posit an eternal and personal God. Scalia cited the law's sponsor in support of this assertion; he referred to "a creator however you define a creator."[1]

In addressing the majority's assertion that the act showed discriminatory preference to creation science, Scalia again cited the legislative history. He stated that the Louisiana legislature sought to achieve a balanced nonindoctrinating curriculum by protecting only those teachers whom they thought were suffering from discrimination; that is, those that taught creation science. He added that the legislators were aware that the court in *Epperson* had already prohibited discrimination against evolution. Scalia observed that since witnesses had informed legislators that most scientists and educators had hostility toward creation science, that creation science had been censored and badly misrepresented, and that there was an unavailability of materials on creation science suitable for classroom use; it was reasonable for the legislators to provide in the act for development of a curriculum guide on creation science, but not on evolution.[1]

Scalia believed that the legislative history of the act gave ample evidence of the sincerity of the Balanced Treatment Act's purpose. Witness after witness, he asserted, urged the legislators to support the act so that students would not be indoctrinated but would instead be free to decide for themselves, based upon a fair presentation of the scientific evidence, about the origin of life. Scalia stated that the legislators made comments during the hearings that showed their sincerity toward the act's articulated purpose, showing a concern about censorship and misrepresentation of scientific information about evolution. Scalia conceded, however, that an awareness of the tension between evolution and the religious beliefs of many children undoubtedly prompted the legislature. However, he added, a valid secular purpose is not rendered impermissible simply because its pursuit is prompted by concern for religious sensitivities.[1]

In summary, Scalia stated that the people of Louisiana are entitled, as a secular matter, to have whatever evidence there may be against evolution presented in their schools. Perhaps, he asserted, what the legislature did was unconstitutional because there is no such evidence, and the scheme that they established would amount to no more than a presentation of the *Genesis* story of creation. However, he stated, this remained to be determined at a trial; it should not have been decided by summary judgment. Personally, he doubted that the scientific evidence for evolution is so conclusive that no one could be "gullible" enough to believe that there is any real scientific evidence to the contrary. Further, he stated that even if the legislature's purpose was to advance religion, some of the well-established exceptions to the impermissibility of that purpose might be applicable if the validating intent was to eliminate a perceived discrimination against a particular religion, to facilitate its free exercise, or to accommodate it. He criticized the court's decision as "*Scopes*-in-reverse."[1]

Both the *McLean* and *Edwards* cases occurred within a time period characterized by increased creationist lobbying in the state legislatures for equal time legislation. From 1980 to 1985, such bills were introduced into nineteen state legislatures.[12] Arkansas and Louisiana were two states that enacted the legislation into law.

The Louisiana Creationism Act involved the Supreme Court of the United States in a difficult constitutional question. The issue framed by the court in *Edwards* was a narrow one, whether the act violated the Establishment Clause of the First Amendment in the special context of the public-school system. Remember, the court historically has given public schools considerable discretion in developing curricula as they see fit. Despite this, the court has been particularly vigilant in monitoring compliance with the Establishment Clause, often invalidating various types of statutes that advance religion in public schools.[1]

The test for Establishment Clause validity at the time of the *Edwards* litigation was the *Lemon* test. Under *Lemon*, the purpose of a legislature in enacting an act into law receives particular consideration and is considered the test's first prong. This is a complex area of the law as evidenced by the different positions the Supreme Court majority and dissent took on this matter in *Edwards*. Analysis shows that two distinct categories of Establishment Clause cases have been identified by the Supreme Court. The first category consists of cases dealing with religious activities within the public schools (the category the court applied to the Creationism Act). The second category involves public aid in varying forms to religious educational institutions. The court has basically handled these two types of cases very differently.[6]

The court has scrutinized purpose far more closely in cases involving religious activities within the public schools.[6] When Scalia, in his dissent, pointed out that before *Edwards* the Supreme Court held state actions unconstitutional for lack of secular purpose in only three other cases, while listing a long line of purpose cases where state actions were held constitutional, he failed to distinguish the two categories of cases.[6] Each of the three cases mentioned by Scalia fell into the category of religious activities within public schools, rather than the category of aid to religious institutions. Scalia basically ignored this distinction by incorrectly lumping together all of the school aid cases with the religion in public school cases[13], thereby minimizing the level of scrutiny applied by the Supreme Court to the latter type of cases. Scalia had to have realized this!

Scalia also critiqued the court majority's contention that the Louisiana Act was unconstitutional because its primary purpose was to endorse a particular religious doctrine.[1] In support of his critique, he cited precedent from certain prior Establishment Clause Supreme Court cases. Scalia maintained that such precedent requires a statute to be wholly or entirely religious in purpose before it can be invalidated. He maintained that since the Creationist Act's stated purpose was the secular purpose of protecting academic freedom, and in his opinion the legislative history of the act seemed to provide at least limited evidence of this, the act should survive under the first prong of the *Lemon* test.[14] However, Scalia made clear that, even if this were not so, he believed that it is not constitutionally permissible to invalidate any act under the Establishment Clause solely on the basis of the purported motivation of the legislature, with no regard to effect. In taking this latter position, he also ignored an entire line of Supreme Court decisions on the subject of purpose and effect.

In sharp disagreement with Scalia, after reviewing the act's legislative history, the majority in *Edwards* rejected the Louisiana legislature's stated purpose of protecting academic freedom as a sham, maintaining that the statute did not and could not serve this purpose. Rather, the majority believed that it attempted to discredit evolution. The majority asserted that the act served to restructure the public-school science curriculum, selecting out for special treatment one scientific theory, evolution, which has historically been disfavored by certain religious sects. The majority, dismissing any scientific validity for creation science, concluded that it embraced a religious doctrine and included a belief in the existence of a supernatural creator. The majority believed that the act's preeminent purpose was to advance a religious point of view or to prohibit evolution from being taught. Either purpose, the majority held, under *Lemon*, is an unconstitutional violation of the Establishment Clause. Therefore, the majority affirmed the summary judgment invalidating the act using this level of scrutiny.

Independent of these substantive considerations was an important procedural one; that is, should a decision have been made by summary judgment rather than after an evidentiary trial on the merits? Under the Federal Rules of Civil Procedure, a summary judgment should not be entered unless there is no genuine issue of material fact to be disputed at an evidentiary trial. Scalia maintained that this was not the situation in *Edwards*. By affirming the summary judgment, Scalia believed that the Supreme Court majority was concluding without a trial that creation science was a religious doctrine rather than a scientific theory.[14] Scalia believed that this was a legal error. He asserted that the court should have limited its ruling to merely whether the act was invalid under the secular purpose prong of the *Lemon* test, without deciding the scientific validity of creation science.[14] Scalia believed that had the court done so, the existence of secular purpose of the statute would have been entirely clear. An analysis by the court of the scientific status of creation science, he believed, should have waited until the completion of an evidentiary trial. He indicated that the parties were sharply divided over the scientific character of creation science. One side asserted that it is a thinly veiled religion doctrine, while the other side asserted that it is a collection of valuable scientific data that can be presented without reference to religion. Scalia believed that support for both interpretations could be found in the legislative history. Scalia, himself, believed that there might be scientific evidence against evolutionary theory and in support of creation science. For example, he stated, "Infinitely less can we say that (or should we say) that the scientific evidence for evolution is so conclusive that no one would be gullible even to believe that there is any real scientific evidence to the contrary, so that the legislature's purpose must be a lie."[1] Scalia apparently found the impossible creationist argument—that there are only two scientific explanations for origins, creation science and evolution—sincere and legitimate.

However, the majority thought otherwise. The *Edwards* court majority believed that creation science is not legitimate science and that the secular purpose necessary to pass constitutional muster under Establishment Clause was nonexistent. The majority's opinion relied in large part on the historical and contemporary link between creation science and fundamentalism and the reliance of creation science on supernaturalism; that is, its dependence on a creator.

Technically, Scalia was correct in asserting that the legislative history of the particular Louisiana statute in question could be interpreted in different ways, and clear testimony and affidavits were presented by certain experts that creation science is not attached to religious doctrine. Nothing on the face of the Louisiana Statute suggested religious doctrine, except perhaps the use of the term creation. The Louisiana statute did not define creation science or set forth its characteristics as did the Arkansas Statute litigated in *McLean*. However, the Louisiana statute litigated in *Edwards* did set forth a two-model approach in similar fashion to the Arkansas Statute. This two-model approach, historically espoused by creationist literature, hypothesized that there are only two possible explanations for earth's life history: creationism or evolution. Under this approach, any evidence against one model functionally serves as evidence in support of the other. Historically, the evolutionary component is defined by creationist literature in traditional Darwinian mechanistic terms; that it, a reliance on natural selection as the major driving force in evolutionary change. Both the *McLean* court and *Edwards* majority constitutionally rejected the two-model approach as motivated by religious purpose. In doing so, they accepted the evidence presented by the mainstream scientific community and their own logical analysis and judicial notice that the two-model system is logically and functionally flawed.[1]

To accept Scalia's position on this matter, one has to ignore the history of creationism and creation science in the United States and the attempts of creationists to insert it into the United States science curriculum. Further, one has to ignore the many alternatives to Darwinism and creationism that have historically been espoused by various scientists and religious. Since the 1970s, many scientists were particularly active in the development of alternative evolutionary theories and mechanisms. These alternatives have generated extraordinary and intense debate within the scientific community, and clearly some are pseudoscience. Alternatives include punctuated equilibrium, pangenesis, Gaia theory, saltation theory, neutral theory, and others. However, none of these alternative theories reject naturalistic mechanisms or interpretations or espouse supernatural ones. In addition, there are even several alternative supernatural theories to creation and evolution, some ancient, including many associated with Eastern religion and philosophy. The two-model approach is just not sustainable when scrutinized objectively.

In addition, mainstream scientific views further convinced the *Edwards* court majority that no real scientific evidence could legitimate a creationism model because of its ultimate reliance on nonnaturalistic, that is, supernatural explanations. The majority accepted this position without question and expected that others of "honest" intent would do the same. Therefore, for purposes of the first prong of the *Lemon* test, the court majority held that the statute must fail for lack of secular purpose. In doing so, the court rejected any attempt to limit the teaching of evolution in the public-school system, either by banning it entirely, or by attaching the teaching of creationism or creation science as a condition to evolution being taught.[1]

Once again, the majority decision in *Edwards*, as did the *Epperson*, *Wright*, and *McLean* decisions, reinforces the presumption of legitimacy that federal courts have afforded mainstream scientists and philosophers, espousing the scientific nature of evolutionary theory, while basically disregarding the testimony of creationists and their expert witnesses as religiously motivated, even when the experts are well-credentialed in secular fields of study. As much as it is clear from the legislative history of the act and the history of creationist attempts at limiting the teaching of evolution in public schools through legislation, it is equally clear that the Supreme Court in this case continued the bias shown by previous federal courts in accepting, *carte blanche,* views of mainstream science and testimony of mainstream experts without any question or skepticism.

Problems exist in justifying this part of the majority opinion. Although nowhere near as broadly or explicitly as was done in *McLean,* the majority entered into the difficult philosophical terrain of defining which elements constitute legitimate science. Ironically, in doing so, they also accepted a two-model approach, albeit of different type than the creationists. The basic philosophical and ideological underpinnings of the majority caused it to view the controversy also in the dualistic terms, evolutionism (naturalism) that, at least in its late twentieth century and early twenty-first century form, encompasses a materialistic view of the universe versus supernaturalism. The majority viewed creationism's ultimate and apparently almost complete reliance on the supernatural, and its undisputable historical Fundamentalist ties as containing fatal religious elements for Establishment Clause purposes. This may be a very legitimate and defensible constitutional approach and solve the problem of giving creation science a place in the public-school science curriculum. However, to take it a step further and enter the difficult terrain of defining science creates problems. Basically, the majority philosophically adopted the position that legitimate science can only be naturalistic, and this became the most important basis of the decision. In effect, the majority believed that no reasonable scientist or person could consider creationism science because of its supernatural underpinning; therefore, the statute must have been motivated by default by religious purpose. Although the *Edwards* Supreme Court majority took a much more judicially constrained position on this issue than did the *McLean* District Court, their basic conclusions in this regard were the same and demonstrated the same bias. It ignored, as did *McLean,* an entire body of work by philosophers of science, discussed in detail in the previous chapter, on demarcation of science from nonscience. The majority position lacked a nuanced and legitimate position in this regard. In this way, Scalia's position on summary judgment is defensible. Scalia urged the majority, in deciding whether a summary judgment should be affirmed, to avoid entanglement in the scientific and philosophical argument of whether scientific creationism is science without, at minimum, an evidentiary trial; rather, he urged the majority to confine itself to a principled look at the purpose of the Louisiana legislature in adopting the statute.

The *Edwards* majority and the *Edwards* dissent in arriving at their respective positions relied heavily on the briefs presented to the court by the involved parties and interested individuals and organizations. A number of legal briefs were filed on behalf of both the appellants and appellees in the *Edwards* case in order to influence the justices. For the most part, the court majority adopted positions espoused in certain appellee briefs. Briefs important in this regard included an *amici curiae* brief filed on behalf of eight scientific and educational liberties groups and seventeen individual scientists, educational, and religious leaders; an *amici curiae* brief filed on behalf of seventy-two Nobel laureates, seventeen state academies of science, and seven scientific organizations; an *amici curiae* brief filed on behalf of Americans United for Separation of Church and State, and other Jewish and Christian religious groups; an *amici curiae* brief filed on behalf of the National Academy of Science, American Jewish Congress and the Synagogue Council of America; and an *amici curiae* brief filed on behalf of the Anti-Defamation League of the B'Nai B'rith and Americans for Religious Liberty. All of these briefs urged the Supreme Court to affirm the district court and Appeals Court summary judgment. Further, all of these briefs maintained that the primary purpose of the act was to advance a particular religious belief and that the act singles out evolution among the sciences for prejudicial treatment.[17, 18, 19, 20] A number of these briefs, particularly the one on behalf of the Nobel laureates and others; the one on behalf of the Natural Academy of Sciences; and the one on behalf of eight scientific and educational liberties groups and seventeen individual scientists, educational, and religious leaders stressed an argument of the type stated in *McLean*; that is, science has certain characteristics, such as it is guided by natural law, it has to be explanatory by reference to natural law, it is testable against the empirical world, its conclusions are tentative, and it is falsifiable. All of these briefs argued that creation science fails these stated criteria and therefore is simply not science; rather, it is an extension of Fundamentalists' religious doctrine. As a result, these briefs urged the Supreme Court to hold that the Louisiana Balanced Treatment Act violated the Establishment Clause, as tested by the *Lemon* analysis.[17, 18, 19, 20]

The appellant/creationist position was also supported by a number of briefs. Among them was a brief filed on behalf of the State of Louisiana; a brief filed on behalf of the Catholic League for Civil and Religious Rights; and a brief filed on behalf of the Rabbinical Alliance of America, the Catholic Center, the Free Methodist Church of America, and a number individual United States Congressmen and the Committee on Openness in Science, a group representing a considerable number of scientists with advanced degrees.[17, 18, 19, 20, 21]

The brief on behalf of the Catholic League and the brief on behalf of the State of Louisiana rejected any assertion that the act violated the Establishment Clause, as tested by *Lemon*. The Catholic League argued that legal precedent holds that the purpose prong of the *Lemon* test yields an Establishment Clause violation only if a law is entirely motivated by a purpose to advance religion. In effect, the league argued that the Appeals Court panel opinion was erroneous to the extent that it required a preeminent secular purpose for the act, when any secular purpose is all that is necessary, in their view, for an act to pass constitutional muster under a purpose test. That is, to fail to have a secular purpose under *Lemon* criteria, a law must be entirely motivated by a purpose to advance religion.[21] This, fundamentally, was the position adopted by Scalia but rejected by the *Edwards* court majority.

The State of Louisiana Brief was an extremely long, detailed document that took special issue with the problem of demarcation of science from nonscience. The Louisiana Brief argued that the *McLean* requirements for what constitutes science are rejected by a majority of philosophers of science. Further, it maintained that among philosophers of science there is little agreement concerning what constitutes science or what criteria are needed for an idea to rise to the level of a scientific theory. The brief cited the philosophers Lakatos and Laudan to make this point. For example, quoting Laudan in his work the *Demise of the Demarcation Problem*, Louisiana maintained that "For most of this century numerous scientists and some philosophers have acted, especially in their more public pronouncements, as if there were clear conceptions of the 'scientific' and the 'pseudoscientific.' Various forms of intellectual activity…have been repeatedly labeled by many scientists and philosophers as 'unscientific' or 'pseudoscientific.' Such pejorative accusations have generally been issued with a confidence and an air of authority that is completely out of character with the murkiness which actually afflicts our conception of the 'scientific.' To further illustrate the problem, the Louisiana Brief quoting Thomas Kuhn maintained that the lines of demarcation themselves shift. For example, Kuhn stated, "The reception of a new paradigm often necessitates a redefinition of the corresponding science. Some old problems may be relegated to another science or declared entirely unscientific." Others that were previously nonexistent or trivial may, with a new paradigm, become the very archetypes of significant scientific achievement. And as the problems change, so often does the standard that distinguishes a real scientific solution from a mere metaphysical speculation, word game, or mathematical play.[17, 18, 19, 20]

The State of Louisiana Brief maintained that in considering the various proposed definitions of science, no group of philosophers of science has formulated a definition that even remotely resembles the definition set forth in *McLean*. Further, it stated that no group of philosophers has subsequently endorsed anything remotely resembling the *McLean* definition, except the ACLU expert witnesses. The Louisiana Brief asserted that the Supreme Court should not restrict the definition of science to this viewpoint.[17, 18, 19, 20]

The Louisiana Brief further contended that there are a variety of definitions of science, with the *McLean* view a positivist and materialist minority approach, one that limits science to natural laws and insists on verification (testing and falsifiability) of all events in order to be scientific. It stated that the basic principle of logical positivism, that all meaningful scientific statements are verifiable by sense experience, is itself a metaphysical principle that cannot be verified by sense experience. Further, the Louisiana Brief argued that modern science itself not only arose from a nonnaturalistic foundation, but also actually depended for its rise on the belief in a creator as a basis for an orderly universe that could offer natural laws.[17, 18, 19, 20]

However, their arguments were to no avail. They were rejected by the *Edwards* majority as they were ignored by the *McLean* District Court some years before. The majority decision in *Edwards* precluded states from mandating a balanced treatment of creation science with evolution in the public-school curriculum. The court majority justified its decision on the basis that the Louisiana Statute was passed with the purpose of endorsing religion; therefore, it failed the first prong of the *Lemon* test. Analytically, the court majority rejected creation science as legitimate science, while accepting evolutionary theory as science. The basis for this distinction seems to be the court's belief that only theories that rely on naturalistic explanation deserve the status of science. The court believed that creation science relies historically and by definition on supernaturalism. This characteristic of creation science proved fatal to the Louisiana Statute and by extension to balanced treatment statutes generally. After (*Edwards*), such statutes were effectively dead.[1]

Although legal opposition to evolution is often legitimately attributed to Fundamentalist activity, it must be acknowledged that widespread belief in evolution is not prevalent in the United States. Even today, many remain skeptical about evolution. A recent survey showed that 44 percent of United States adults accept a literal *Genesis* account of creation, rather than evolution. A 1993 survey showed that 47 percent of adult United States citizens believe that humans were created by God less than ten thousand years ago. More recent polls show much the same results, with a substantial majority stating that they want creationism taught with evolution in public school.[22, 23, 25, 26]

These surveys illustrate the conceptual distance between the general lay population in the United States, almost half of whom reject evolutionary theory, and the professional scientific community, who overwhelmingly accepted some form of evolution since the late nineteenth century.

Exacerbating this problem is the way evolutionary theory is understood today by many biologists. A rejection of any type of progressive evolution or teleology in nature is common. For example, Gould, in espousing this position, spared no disdain for those that

92

hold a literal interpretation of the *Genesis* creation story or for those that believe evolution is directed toward a goal (such leading ultimately toward humans, either because of the supernatural or otherwise). He wrote, "all thinking people accept the biological fact" of evolution. Gould asserted that life on earth is not the result of special creation, nor is it the result of progressive or teleological processes.[24]

However, despite their notable success in the federal courts with the vicious attacks leveled against creationism, those that have resisted creationist attempts to eliminate or limit evolution's place in the public-school curriculum have not been successful in ending Fundamentalist legal challenges to an evolutionary monopolized science curriculum. Nor have they stopped biblical literalists from seeking alternative formats in the public schools for discussion of creationist alternatives. Therefore, this peculiar legal battle continued into the twenty-first century in a new incarnation of scientific creationism, intelligent design, as the next chapter discusses.

INTELLIGENT DESIGN

Kitzmiller v. Dover Area School District

Between the *Edwards*[1] decision in 1987 and *Kitzmiller v. Dover Area School District*[2] decision in 2005, creation science recast itself into a seemingly different form called intelligent design (ID). Proponents of ID believed that its inclusion in the public-school biology curriculum was more likely to survive a First Amendment challenge than its creation-science predecessor. The transition from creation science to intelligent design is a story told in the *Kitzmiller* case.

However, the period between *Edwards* and *Kitzmiller* was also marked by impressive discoveries in the field of evolutionary biology. This was due in large part to advances in the fields of developmental biology, molecular biology, genetics, and the maturation of a new field of study called evolutionary developmental biology or evo-devo. Many questions concerning evolutionary mechanisms, tempo, and mode of change became understood in deeper and more complete ways than ever before.

Darwin used embryology (developmental biology) as evidence for evolution in *On the Origin of Species,* first published in 1859, as did others in various ways after him. Darwin believed that evidence for evolution can be seen from comparing similar embryonic structures (called homologies) among different groups of animals. He also believed that evolutionary significant modifications in form can be seen by observing how development is different or altered in various species. Darwin believed that the emergence of structures that distinguish different types species from one another generally take place later in embryonic development. For example, the similarities in various vertebrate embryos as diverse as birds, fish, whales, and humans are striking in early embryonic development; structures that distinguish these different species from one another emerge somewhat later in their development.[3]

Despite long-running interest by naturalists in the hypothesized link between development and evolution, little was known about the genes important in this link. This situation started to change twenty-five or so years ago as the new field of evolutionary developmental biology, or evo-devo as it is often called, grew from the merger of developmental biology, genetics, and evolution. This merger is producing a new model of evolution, integrating developmental genetics to explain evolutionary change.[3]

Evo-devo is a study of how changes in developmental genes create diverse variations that natural selection can act upon.[3] Evolutionary biologists are discovering how these developmental genes work to change body pattern during development. An important group of these types of genes are the *Hox* genes. *Hox* genes are responsible for what are called homeotic transformations or mutation in animals. We have learned that these genes are regulatory genes, meaning that they control the activity of many other genes and ultimately developmental pathways which are responsible for the formation of body parts in animals.[3] Actually, *Hox* gene clusters and homeotic transformations were first observed in insects, namely in *Drosophila*, a fruit fly genus, in the 1940s by C. H. Waddington and Richard Goldschmidt. They identified mutations where one body segment of the fruit fly was transformed into a different type of segment. They called these "homeotic mutants." Waddington and Goldschmidt believed that these mutants might be a key for understanding the relationship between genetics, development, and evolution. An example of a homeotic mutations is the replacement of fruit fly halteres (balancers) with wings, resulting in a four-winged mutant fly, rather than the typical two-winged type.[3]

In the 1980s and 1990s, *Hox* genes, to the surprise of many investigators, were also found in vertebrates, and they are now thought to be present in all animals, including mammals such as humans. Perhaps, even more shocking, no matter what animal species is looked at, be it fruit flies, mice, or humans, these genes are always found in the same order and sequence, and they have the same basic function. They differ only in their number and number of sets.[3] They even can be functionally interchangeable among certain species like flies and mice. The *Hox* genes are all regulatory genes that code for protein products that control the expression of other genes.

The *Hox* clusters are groups of homeotic genes that function in directing or participating in the embryonic development. They are important in determining developmental structures along the anterior-posterior axis of the embryo in both invertebrates and vertebrates, including humans. Genetic similarities among these various diverse species are often striking. For sexample, the human *Hox B4* gene mimics the function of the *Drosophila* (fruit flies) *deformed* gene when genetically engineered into *Drosophila* embryos.[3]

The *Hox* genes are not the only developmentally significant genes that have been discovered that might have evolutionary significance. There are many other genes of this type found in plants and animals. Another example is the *Pax6* gene involved in eye development in vertebrates. Research in the 1990s showed that genes involved in eye development in different phyla of animals are homologous. It was found that similar aberrant eye development is found in fruit flies, mice, and men, caused by mutations in *Pax6*s, or the invertebrate analogue called *eyeless*. Interestingly, *Pax6* and *eyeless* are also regulatory genes that can be interchanged in different species and still function.[3] However, when any type of regulatory gene is engineered into a donee species, it functions in a way that is specific to the donee species.

The information emerging from evolutionary developmental biology is changing how evolution is viewed mechanistically. For some, it calls into question traditional Darwinian views that evolution proceeds generally in gradual manner. *Hox* gene mutations, for example, can cause discontinuous jumps in morphology of body structure. The evolutionary significance of this is debatable and being debated, but nonetheless compelling and interesting. In addition, evo-devo experiments are providing insights into what developmental constraints operate in organisms to prevent certain evolutionary pathways or trajectories. This allowss nuanced explanations for some common evolutionary patterns, such as stases, convergent evolution, reversibility of certain morphological changes, and the absence of morphological features in particular lineages.[3]

Another field with evolutionary significance that emerged over the last twenty years or so is the possible role of epigenetic inheritance in evolutionary change. Recent evidence suggests that the Darwinian and New Synthesis view that natural selection only operates on variations that have a genetic basis may not be completely correct. It is now known that sometimes variations not due to DNA sequence changes in a wide range of single-celled and multi-celled organisms might be subject to evolutionary selection. This is called epigenetic inheritance. Epigenetic inheritance has several molecular causes that have nothing to do with genetic changes or mutations. The best studied cause is enzymatic methylation of specific regions of the DNA (adding a carbon and hydrogen groups). The methylation process is often repeated or retained over several or many cell divisions or generations, and it has the effect of reducing or eliminating gene activity in that region. Epigenetic inheritance has been associated with many characteristics of many species and it is even responsible for certain "mutant" phenotypes; for example, petal-shaped in the toadflax. In the toadflax, mutant flowers can be produced by one of two ways. They can be produced by changes (mutations) in the DNA sequence of a relevant gene or by extensive methylation of a gene that controls flower symmetry without alteration of the nucleotide sequence. Even certain cancers are the result from epigenetic changes. Environmental factors have been shown to produce epigenetic changes in genomes. It is increasingly believed that if epigenetic changes are heritable and span generations, which some seem to be, natural selection might operate on the variation they produce. Therefore, a modern view of evolution is emerging that includes epigenetics. It has been stated that to understand evolution one must understand how natural selection, genes, and nongenetic inheritance interact.[3]

These are just a few of the many advances that occurred over the relevant years in evolutionary biology as new information was gathered. More will be discussed later in this chapter. These studies have generated hundreds of refereed papers in scientific journals.

Evolution is a vibrant and dynamic intellectual discipline. It is science. Views change as new information is gathered and new techniques to probe the nature of life are developed and utilized. More and more evidence is accumulating that supports evolution as the best explanation for the history of life on earth.

Now with this backdrop on how the study of evolution advanced and increased our understanding of life and how it develops and transmutates, let's compare the development of creation science during the same relevant time period. To do this, let's turn to the *Kitzmiller* decision as it explores how creation science gave birth to intelligent design and how creationists have used this in attempting to circumvent the *Epperson* and *Edwards* decisions.

The *Edwards* case ended any creationist hope that creation science can be a vehicle to balance evolution in the public-school curriculum with creationism, or eliminate evolution from the science curriculum. However, *Edwards* did not end balanced treatment efforts. The fight for balanced treatment continued with creation science recast into what is known as intelligent design (ID). An attempt to limit the teaching of evolution in the public schools with intelligent design was addressed in *Kitzmiller,* a Pennsylvania federal district court case. The case never went beyond that of district court; the decision was not appealed to a higher federal court for reasons that will be discussed, yet the *Kitzmiller* decision is very significant with far-reaching influence.[1, 2]

Tammy Kitzmillar, the mother of ninth and eleventh graders in Dover Area High School, Pennsylvania, at the time the case was initiated, along with a number of organizations, including the ACLU and Americans United for Separation of Church and State, were plaintiffs in the case. The defendants were Dover Area School District and Dover Area School District Board of Directors.[2]

The incidents that led to the *Kitzmiller* case began in October 2004 when the Dover Area School District's Board of Directors passed a resolution stating that "Students will be made aware of gaps/problems in Darwin's theory and of other theories of evolution including, but not limited to, intelligent design." In November of the same year the school district followed with a press release announcing that beginning in January 2005 teachers would be required to read the following statement to students in the ninth-grade biology class at Dover High School: "The Pennsylvania Academic Standards require students to learn about Darwin's Theory of Evolution and eventually to take a standardized test of which evolution is a part. Because Darwin's Theory is a theory, it continues to be tested as new evidence is discovered. The Theory is not a fact. Gaps in the Theory exist for which there is no evidence. A theory is defined as a well-tested explanation that unifies a broad range of observations. Intelligent Design is an explanation of the origin of life that differs from Darwin's view. The reference book, *Of Pandas and People*, is available for students who might be interested in gaining an understanding of what Intelligent Design actually involves. With respect to any theory, students are encouraged to keep an open mind. The school leaves the discussion of the Origins of Life, to individual students and their families. As a Standards-driven district, class instruction focuses upon preparing students to achieve proficiency on Standards-based assessments."[2]

In December 2004, Plaintiffs filed a lawsuit in a Pennsylvania federal district court challenging the constitutionality of the resolution and press release (collectively referred to as "the ID Policy"). They contended that the ID Policy constituted an establishment of religion prohibited by the First Amendment to the United States Constitution and the Constitution of the Commonwealth of Pennsylvania. Plaintiffs asked the court for declaratory judgment and injunctive relief preventing the implementation of the ID Policy.[2]

A trial began in September 2005 and continued through November 2005. The district court issued a long memorandum opinion, after hearing all testimony presented by the parties. In the opinion the court concluded that the ID Policy violated the Establishment Cause of the First Amendment of the United States Constitution and Pennsylvania Constitution.[2]

The basis of the *Kitzmiller* court's decision centered on what constitutes ID and on the purpose and effect of the defendants' actions. After days of testimony during an evidentiary trial, the court linked intelligent design to fundamentalism and labeled it the religious heir to creation science. The court stated that the ID Policy was the School District and Board's attempt to reintroduce, in a different form, "balanced treatment" of creationism and evolution into the public-school science curriculum.[2]

As *McLean* and *Edwards* did before it, *Kitzmiller* used the *Lemon* test to evaluate the constitutionality of the ID Policy. However, for the first time in evolution litigation, it also used a newer constitutional test, the so-called endorsement test, in its analysis. Although endorsement language was used in a number of prior cases, including *Edwards*, an endorsement analysis was used explicitly for one of the first times by the Supreme Court in a concurring opinion in 1989 in *County of Allegheny v. ACLU*, after *Edwards* was decided.[1, 4, 5, 6, 7]

The *Kitzmiller* court believed that the endorsement test was relevant because the central issue in the case was whether the government endorsed religion by its ID Policy. The endorsement test relies on a hypothetical "reasonable observer," defined as an informed citizen who is more knowledgeable than the average person about the relevant issues of the case. However, I must add at this point that the endorsement test standard, at the time *Kitzmiller* was decided, was not as clearly defined as the *Kitzmiller* court led us to believe. For example, in criticism of the reasonable observer standard used by *Kitzmiller* and other courts, it has been stated, "ultimately leaving us with the uncomfortable inclination that this "purely fictitious character will perceive precisely as much, and only as much, as its author wants it to perceive."[8]

In applying constitutional tests, the *Kitzmiller* court reviewed the history of the ID movement. An issue driving *Kitzmiller* was whether intelligent design (ID) is science, or it is just another form of creationism; that is, religion in a different guise. In attempting to understand exactly what ID is, *Kitzmiller* addressed its religious roots and character.[2]

The court related that the plaintiff witness, Dr. John Haught, a theologian and author who has written extensively on evolution and religion, testified that ID is not a new scientific hypothesis or theory; rather, he stated that it is an old religious argument for the existence of God. Haught traced ID's origin back to Thomas Aquinas in the thirteenth century, who wrote that, wherever complex design exists, there must be a designer. Haught testified that Aquinas was explicit that this intelligent designer was the Judeo-Christian God. The *Kitzmiller* court concluded that this is essentially the same argument for ID that was presented by defense expert witnesses (Professors Behe and Minnich, both scientists). Haught further testified that this was also an argument for the existence of God, advanced in the nineteenth century by Reverend Paley. Behe and Minnich actually admitted that ID's argument that complex organisms show "purposeful arrangement of parts" is the same one that Paley made for design. The only apparent difference between the argument made by Paley and the argument for ID, as expressed by defense expert witnesses, is that ID's "official position" does not acknowledge that the designer is God. However, the court continued, Haught testified that anyone familiar with Western religious philosophy would immediately make the association that the unnamed designer is indeed God.[2]

The *Kitzmiller* court stated that Haught's position is supported by the description of the designer in *Of Pandas and People,* an important ID textbook mentioned in Dover's ID Policy, as a "master intellect," suggesting a supernatural deity as the designer. The court concluded that this is even acknowledged in *Pandas* with its statement, "what kind of intelligent agent was it [the designer]" and answer: "On its own science cannot answer this question. It must leave it to religion and philosophy." *Pandas* also claims that there are two kinds of causes, natural and intelligent. Intelligent causes are beyond nature. Haught, who was the only theologian to testify in this case, explained that in Western intellectual tradition, nonnatural causes occupy a space reserved for ultimate religious explanations. The court significantly noted that not one defense expert was able to explain how the supernatural action suggested by ID could be anything other than an inherently religious proposition.[2, 11]

The *Kitzmiller* court also recounted that Dr. Barbara Forrest, one of the plaintiffs' expert witnesses, the author of the book *Creationism's Trojan Horse*[12], testified that ID is religious in nature. She based this on statements of many of its proponents. For example, she recounted that Phillip Johnson, a leading ID leader and author of the 1991 book titled *Darwin on Trial*,[13] wrote "that God is objectively real as Creator and recorded in the biological evidence…" Johnson concluded that science must be redefined to include the supernatural if religious challenges to evolution are going to get a hearing.[2, 12, 13]

Kitzmiller stated that the defendants' expert witnesses, Dr. Michael Behe and Dr. Scott Minnich, confirmed that the existence of a supernatural designer is a hallmark of ID. For example, Behe wrote that by ID he means "not designed by the laws of nature," and that it is "implausible that the designer is a natural entity." Moreover, the court stated that Behe admitted that his personal view is that the designer is God. Minnich also testified that the designer is God and that many leading advocates of ID also believe this. He testified that for ID to be considered science, the ground rules of science have to be broadened so that supernatural forces can be considered.[2]

Kitzmiller also concluded that evidence of ID's religious nature is found in the "Wedge Document," developed by the Discovery Institute's Center for Renewal of Science and Culture (hereinafter "CRSC"), a conservative organization that advocates ID. This document stated in its "Five Year Strategic Plan Summary" that its goal was to replace science as currently practiced with "theistic and Christian science." As posited in the Wedge Document, the Institute's "Governing Goals" were to "defeat scientific materialism and its destructive moral, cultural, and political legacies" and "to replace materialistic explanations with the theistic understanding that nature and human beings are created by God." The CSRC expressly announced in the Wedge Document a program of Christian apologetics to promote ID. A careful review of the Wedge Document's goals and language revealed to the court the CSRC's cultural and religious goals, as opposed to scientific ones. For example, the court related that ID aspires to change the ground rules of science to make room for religion, specifically, beliefs consistent with a particular version of Christianity.[2]

After placing the legal challenge to the teaching of evolution and ID in historical context, *Kitzmiller* turned its attention to testing the ID policies constitutional viability using the "hypothetical reasonable observer" standard of the endorsement test. The court stated that an adult or child, who is "aware of the history and context of the community and forum" is also presumed to know that ID is a form of creationism. The court asserted that the evidence at trial demonstrated that ID is nothing more than the offspring of creationism.[2]

Kitzmiller, in applying the endorsement test, stated that the strongest evidence supporting the finding of ID's creationist nature is the history of the book, which students in Dover's ninth-grade biology class were referred, *Of Pandas and People. Pandas* was published by the Foundation for Thought and Ethics (FTE), an organization whose articles of incorporation and filings with the Internal Revenue Service describe it as a religious, Christian organization. *Pandas* was written by Dean Kenyon and Percival Davis, both acknowledged creationists, and Nancy Pearcey, a young earth creationist, contributed to the work.[2, 11]

Plaintiffs' witnesses testified that *Pandas* went through several drafts, some of which were completed prior to, and some after, the Supreme Court's decision in *Edward.* By comparing the pre- and post-*Edwards* drafts of *Pandas,* the court found that the definition for creation science in early drafts was identical to the definition for ID in later ones. However, noteworthy after *Edwards* held that creation science was religion, the words *creation science* and *scientific creationism* were replaced by intelligent design. *Kitzmiller* believed that this was compelling evidence supporting plaintiffs' assertion that ID is creationism relabeled and that the objective observer, whether adult or child, would conclude from *Pandas* that the intelligent designer is God.[2, 11]

Further evidence given by the court to support the conclusion that a reasonable observer, adult or child, would know that ID is a form of creationism was from the testimony of Dr. Forrest, a plaintiffs' witness. Forrest stated that there are six arguments common to ID and creationism that support the premise that they are the same philosophy: the rejection of naturalism, evolution's alleged threat to culture and society, "abrupt appearance" implying divine creation, the alleged gaps in the fossil record, the alleged inability of science to explain complex biological information like DNA, and the dualism that creationism or ID is the only alternative to evolution.[2]

The *Kitzmiller* court did acknowledge that defense experts Behe and Minnich testified that ID is not creationism; however, the court stated that their testimony failed to rebut the creationist history of *Pandas* or other evidence presented by plaintiffs showing the commonality between creationism and ID.[2]

The *Kitzmiller* court believed that the Dover Board and School District's actions and disclaimer also failed to meet the endorsement test for both adults and students. The *Kitzmiller* court, citing a standard articulated in *Edwards,* stated that courts must be particularly "vigilant in monitoring compliance with the Establishment Clause in elementary and secondary schools because they are very impressionable, and their attendance is involuntary.[2]

In ascertaining whether an objective Dover High School ninth grade student would view the disclaimer as an official endorsement of religion, *Kitzmiller* stated that a reasonable, objective student is not a specific, actual student or even an amalgam of actual students but is instead a hypothetical student, one to whom the reviewing court imputes detailed historical and background knowledge, and also one who interprets the challenged conduct in light of that knowledge with the level of intellectual sophistication that a child of the relevant age would bring to bear.[2]

Using this standard, *Kitzmiller* concluded that an objective student would view the disclaimer's plain language as a strong official endorsement of religion. In addition to what has been discussed, the court also based its conclusion on several factors, including a paragraph-by-paragraph analysis of the disclaimer. For example, evolution is treated differently by defendants than all other science subjects. By presenting ID as an alternative to evolution, the students are confronted with the same "contrived dualism" that the *McLean* court recognized to be a creationist tactic with "no scientific factual basis or legitimate educational purpose," but merely religious.[2]

In essence, *Kitzmiller* concluded that the Dover defendants' ID Policy, with its disclaimer, singles out the theory of evolution for special treatment, misrepresents its status in the scientific community, causes students to doubt its validity without scientific justification, presents students with a religious alternative masquerading as a scientific theory, directs them to consult a creationist text as though it were a science resource, and instructs students to forgo scientific inquiry in the public-school classroom. Furthermore, the court stated that introducing ID necessarily invites religion into the science classroom as it sets up what would be perceived by students as a "God-friendly" science, one that explicitly mentions an intelligent designer, and that the "other science," evolution, takes no position on religion. This was supported by plaintiff experts who testified that a false duality is produced that "tells students…quite explicitly, choose God on the side of intelligent design or choose atheism on the side of science." The court believed that introducing such a religious conflict into the classroom is "very dangerous" because it forces students to "choose between God and science," not a choice that schools should be forcing on them. The court stated that an objective student would view the disclaimer as a strong official endorsement of religion.[2]

I find the court's view on this matter interesting, and perhaps somewhat disingenuous. All of the things stated by the court may very well be objectively correct; however, this so-called "hypothetical," "objective" student was a nothing more than a subjective construct contrived by the court as a vehicle to express views made by an adult judge.[8]

The court next addressed specifically whether an objective adult observer in the Dover community would perceive defendants' conduct similarly. The *Kitzmiller* court concluded that an objective adult observer in the Dover community would also perceive the ID Policy as an endorsement of religion for reasons already stated and for additional reasons. For example, the board implemented its curriculum changes publicly, thus reaching out to the entire community as the "listening audience" for its religious message. This was done in a number of ways, including public-school board meetings where the curriculum change was discussed in religious terms; through a newsletter sent to community households denigrating evolution while advocating ID; and through a letter sent to parents of children taking biology "asking if anyone ha[s] a problem with the [disclaimer] statement," and calling on them to decide whether to allow their children to remain in the classroom and hear the religious message or instead to direct their children to leave the room. In addition, the court believed that an objective adult member of the Dover community would know that ID and teaching about supposed gaps and problems in evolutionary theory are creationist religious strategies that developed from earlier forms of creationism, as does the disclaimer's declaration that evolution "is a theory...not a fact." *Kitzmiller* concluded that the objective observer is therefore aware of the social context in which the ID Policy arose, and considered in light of this history, the challenged ID Policy constitutes an endorsement of a religious view.[2]

After concluding that the ID Policy failed to meet the requirements of the endorsement test, thereby violating the Establishment Clause of the First Amendment to the federal constitution, the *Kitzmiller* court could have ruled in favor of the plaintiffs without need for any further review or analysis. However, the court was apparently on a mission to destroy the legitimacy of ID as science completely, irrevocably, and without redemption, just as *McLean* attempted to do with creation science. To complete the overkill, the court turned to a *Lemon* test analysis.[2, 5]

The *Lemon* test is also a traditional test for Establishment Clause violations. As has been discussed in previous chapters, it was first set forth by the Supreme Court of the United States in *Lemon v. Kurtzman* in 1971. The court in *Lemon* proposed three criteria or prongs to test a statute for Establishment Clause violation. First, the statute in question must have a secular purpose. Second, its principal or primary effect of the statute must be one that neither advances nor inhibits religion. Third, the statute must not foster an excessive entanglement with religion. A statute violates the Establishment Clause if it fails to satisfy even one of these three prongs. The *Lemon* test was first used in evolution in public school cases in *Daniel v. Waters* in the 1970s, and it has been used by federal courts in every case of this type since *Daniel*.[2, 5, 10]

In applying the *Lemon* test, the *Kitzmiller* court's main focus was "whether the School District showed favoritism toward religion generally or any set of religious beliefs in particular," stating that the "First Amendment mandates governmental neutrality between religion and religion, and between religion and nonreligion. When the government acts with the ostensible and predominant purpose of advancing religion, it violates the central Establishment Clause value of official religious neutrality, there being no neutrality when the government's ostensible object is to take sides." The court continued *Lemon's* purpose prong "asks whether government's actual purpose is to endorse or disapprove of religion. A governmental intention to promote religion is clear when the State enacts a law to serve a religious purpose." *Kitzmiller* looked at the actual language of the ID Policy within its historical context to determine whether it served to endorse religion.[2, 5]

The court stated that the disclaimer's plain language, the legislative history, and the historical milieu in which the ID Policy arose, led it to the conclusion that the defendants consciously chose to change Dover's biology curriculum to advance creationism, which *Edwards, McLean*, and other courts have held to be religion.[1, 2, 10]

As evidence for its conclusion, the court cited a chronology of the events from as early as 2002, which culminated in enactment of the ID Policy. These included board members stating that "creationism" is an important issue and that it "belonged in biology class alongside evolution." In addition, evolution was disparaged by board members on religious grounds, and board members met with Dover science teachers and expressed concern that evolution was being taught as fact rather than "theory." *Kitzmiller* noted that nothing like this ever happened before in the school district concerning any other subject, and the court felt that these actions caused a chilling effect on teaching evolution.[2]

In addition, the school board delayed purchasing the biology textbook recommended by the biology faculty and administration because the book treated evolution as fact, and it did not discuss any alternatives to the theory of evolution. The teachers were provided a survey of biology books used in private religious schools in York County and a product profile of a biology textbook used at Bob Jones University (a Fundamentalist university). The board stated that it would only approve the textbook chosen by faculty if *Pandas* was also purchased as a supplemental textbook. Incredibly, the board solicitor warned the board that its actions might cause litigation because of past court decisions; however, his warning was ignored. Throughout the entire process, the teachers' views on the matter were largely ignored. The board relied almost entirely on *Pandas* to assist its decision-making on ID without seeking advice from the National Academy of Science, the American Association for the Advancement of Science, the National Science Teachers' Association, the National Association of Biology Teachers, or any other national organizations concerned with science education. Yet as expert testimony indicated, all of these organizations have information about teaching evolution readily available on the Internet, and they include statements opposing the teaching of ID.[2]

Further, the teachers opposed the board's disclaimer in a memo and requested they not be made to read it. The memo sent to the board stated, "Intelligent design is not science, intelligent design is not biology. Intelligent design is not accepted as scientific theory." However, despite this, the board ignored the teachers.[2]

The court concluded that these actions and other like actions by the board made it clear that ID Policy did not have a secular purpose mandated by the First Amendment to the federal constitution. The *Kitzmiller* court, citing *Edwards*, also concluded that although courts are "normally deferential to a State's articulation of a secular purpose, it is required that the statement of such purpose be sincere and not a sham." *Kitzmiller* stated that although the defendants consistently asserted that the ID Policy was enacted for the secular purposes of improving science education and encouraging students to exercise critical thinking skills, the board took none of the steps that school officials would take if these stated goals had truly been their objective. As indicated, the board consulted no scientific materials on pertinent subjects. The board contacted no scientists or scientific organizations. The board failed to consider the views of the district's science teachers. The board relied solely on legal advice from two organizations with demonstrably religious, cultural, and legal missions, such as the Discovery Institute. Moreover, defendants asserted that secular purpose of improving science education is contradicted by the fact that most, if not all, of the board members who voted in favor of the biology-curriculum change conceded that they still do not know, nor have they ever known, precisely what ID is.[1, 2] *Kitzmiller* concluded that to claim a secular purpose against this history is not credible. Therefore, the court stated that defendants' actions failed to pass constitutional muster under the purpose prong of *Lemon*.

Once again, at this point, the court needed no analysis to reach a final decision, yet it continued to further address constitutional issues.[2, 5] The *Kitzmiller* court next turned its attention to the effect prong of the *Lemon* test. Citing the Supreme Court of the United States in the 1989, *Texas Monthly, Inc. v. Bullock*, case, *Kitzmiller* stated that the "principal or primary effect" of a governmental action must be one that neither advances nor inhibits religion, "nor be overtly hostile to religion, nor place its prestige, coercive authority, or resources behind a single religious faith or behind religious belief in general, compelling nonadherents to support the practices or prose-lytizing of favored religious organizations and conveying the message that those who do not contribute gladly are less than full members of the community.[2, 5, 14]

Applying this standard, the court concluded that since it did not believe that ID is science, the only real effect of the ID Policy was the advancement of religion. Further, the *Kitzmiller* court stated that the board's disclaimer read to the students had the effect of bolster-ing religious theories of origins by attacking evolution, disavowing endorsement of educational materials on evolution, and advocating alternative religious concepts. The court concluded that the effect of the defendants' actions in adopting the curriculum change was to impose a religious view of biological origins into the biology course in violation of the establishment clause.[2]

At this point, truly nothing further needed to be decided or determined. Nonetheless, the animus of the court toward ID enticed it to enter the difficult, controversial, and complicated area of defin-ing what science is. As discussed in previous chapters, this is rug-ged intellectual terrain that both *McLean* and *Edwards* risked with less-than-successful results.

Kitzmiller concluded that ID is not science for several reasons. First, the court believed that ID violates a vital ground rule of science by relying on supernatural cause; second, the argument of irreducible complexity, central to ID, employs the same flawed and illogical contrived dualism as creation science in the 1980s; and third, ID's negative attacks on evolution have been refuted by the scientific community. In addition, the court noted that ID has failed to gain acceptance in the scientific community: it has not generated peer-reviewed publications, nor has it been the subject of testing and research.[2s]

Relying on plaintiffs' expert testimony, the court recounted that since the scientific revolution of the sixteenth and seventeenth centuries, science has been limited to the search for natural causes using empirical evidence and testability to explain natural phenomena. The court stated that this has often been called "methodological naturalism," or the scientific method, and it is a "ground rule" of science today.[2]

I must add at this point that the court's view in this regard was oversimplified and inaccurate. Even in the nineteenth century, nonnaturalistic explanations for biological phenomena were still held and set forth. Many examples can be given to support this. One that seems appropriate to the issue at hand is found with Alfred Russel Wallace, who independently—but contemporaneous with Darwin—discovered natural selection. Wallace was a hyperselectionist, who saw natural selection as the sole mechanism driving evolutionary change. He criticized Darwin for his unwillingness to also accept this view of evolution. Darwin believed that although natural selection is the most important mechanism of evolutionary change, it is not the only one. As an outgrowth of his hyperselectionist views on evolution, Wallace did not believe that the human brain, intellect, and morality can be a product of natural selection. Wallace saw the brain as vastly overdeveloped for what was needed to survive and as a result beyond the power of natural selection to form. Therefore, since Wallace concluded that natural selection is evolution's only mechanism, he believed that some higher supernatural power must have intervened to construct the brain. Darwin was very distressed by

Wallace's views in this regard. He wrote to Wallace in 1869: "I hope you have not murdered too completely your own and my child." Darwin was referring, of course, to natural selection.[15]

With that aside, back to the *Kitzmiller* decision. *Kitzmiller* continued its assertion that science must only seek and hold naturalistic explanations by citing the position of the National Academy of Sciences (NAS), recognized by the plaintiffs' and defendants' experts as the "most prestigious" scientific association in this country. The NAS stated, "Science is a particular way of knowing about the world. In science, explanations are restricted to those that can be inferred from the confirmable data—the results obtained through observations and experiments that can be substantiated by other scientists. Anything that can be observed or measured is amenable to scientific investigation. Explanations that cannot be based upon empirical evidence are not part of science." The court believed that a rigorous attachment to "natural" explanations is an essential attribute to science by definition and by convention. Continuing to cite the NAS, the court wrote that creationism, intelligent design, and other claims of supernatural intervention in the origin of life or of species are not science because they are not testable by the methods of science.[2]

The court also stated that the American Association for the Advancement of Science (AAAS), the largest organization of scientists in this country, has taken a similar position on ID, namely, that it "has not proposed a scientific means of testing its claims" and that "the lack of scientific warrant for so-called 'intelligent design theory' makes it improper to include as part of science education."[2]

The *Kitzmiller* court noted that not one expert witness over the course of the six-week trial was able to identify even a single major scientific association, society, or organization that endorsed ID as science. The court also stated that even defense experts conceded that ID has achieved no acceptance in the scientific community. The court concluded that ID fails to meet the threshold necessary to be considered science.[2]

The *Kitzmiller* court agreed with plaintiffs expert Dr. Miller that, from a practical perspective, attributing unsolved problems about nature to causes and forces that lie outside the natural world is a "science stopper." Dr. Miller explained that once you attribute a cause to an untestable supernatural force, a proposition that cannot be disproven, there is no reason to continue seeking natural explanations. Relying on testimony from various plaintiffs experts, *Kitzmiller* stated that ID is dependent on supernatural causation; that is, ID takes natural phenomena, and instead of accepting or seeking natural explanations, argues that the explanation is supernatural.[2]

Kitzmiller asserted that further support for the position that ID relies on supernatural, rather than natural causation, is found in the ID reference book, *Pandas*, which ninth-grade biology students were advised to consult by the board and school district. The court quoted *Pandas* to support its view: "Darwinists object to the view of intelligent design because it does not give a natural cause explanation of how the various forms of life started in the first place. Intelligent design means that various forms of life began abruptly, through an intelligent agency, with their distinctive features already intact—fish with fins and scales, birds with feathers, beaks, and wings, etc." The court concluded that ID asserts that animals did not evolve naturally through evolutionary means but were created abruptly by a nonnatural, or supernatural, designer and *Kitzmiller* stated even defendants' expert witnesses, Behe and Fuller, acknowledged this point.[2]

The court noted that ID in reality attempts to broaden the definition of modern science by also recognizing supernatural causation in the material world as well as natural causation. For example, defense expert Professor Fuller testified that ID aspires to "change the ground rules" of science, and lead defense expert Professor Behe admitted that his broadened definition of science, which encompasses ID, would even embrace astrology. Incredibly, defense expert Professor Minnich acknowledged that for ID to be considered science, the ground rules of science have to be broadened to allow consideration of supernatural forces. *Kitzmiller* stated that ID leaders are in agreement with opinions expressed by defense expert witnesses that the ground rules of science must be changed and broadened for ID to be included as mainstream science. The court stated that William Dembski, a well-known ID leader, stated that science is governed by methodological naturalism, and he argued that this must be expanded for ID to be widely accepted.[2]

The *Kitzmiller* court continued that the Discovery Institute, an ID think tank, acknowledges as "governing goals" to "defeat scientific materialism and its destructive moral, cultural and political legacies" and "replace materialistic explanations with the theistic understanding that nature and human beings are created by God." In addition, and as previously noted, the Wedge Document, developed by the Discovery Institute, states in its "Five Year Strategic Plan Summary" that its intent is to replace science as currently practiced with "theistic and Christian science," seeking nothing less than a complete scientific revolution in which ID will supplant evolutionary theory.[2]

The court stated that ID is based upon a false duality; namely, that if evolutionary theory is discredited, ID is confirmed. *Kitzmiller* concluded that this duality was previously dismissed in *McLean* in relations to creation science. The *Kitzmiller* court, quoting *McLean*, noted the "fallacious pedagogy of the two-model approach" and that "[i]n efforts to establish 'evidence' in support of creation science, the defendants relied upon the same false premise as the two-model approach...all evidence which criticized evolutionary theory was proof in support of creation science."[2]

In addition, the *Kitzmiller* court rejected a primary centerpiece argument of ID: that is, the so-called "irreducibly complex" features of certain biochemical systems and anatomical structures of organisms argue against evolution. The court noted that defense expert Behe is a main proponent of this concept. In his book, *Darwin's Black Box* (20), and subsequently modified in his 2001 article titled "*Reply to My Critics*,"[21] Behe discussed what he meant by an irreducibly complex system. He stated that it is a system composed of several well-matched, interacting parts that contribute to the basic function, wherein the removal of any one of the parts causes the system to effectively cease functioning. Behe stated that an irreducibly complex system cannot be produced directly by slight, successive modifications from a precursor system, because any precursor to an irreducibly complex system that is missing a part is by definition nonfunctional. Since natural selection can only choose systems that are already working it would, according to Behe, not be able to produce such a system.[2]

The *Kitzmiller* court rejected this argument for several reasons. The court asserted that Behe, himself, admitted in "*Reply to My Critics*" that there is a defect in his view of irreducible complexity because, while it purports to be a challenge to natural selection, it does not actually address "the task facing natural selection." Professor Behe specifically explained that "[t]he current definition puts the focus on removing a part from an already-functioning system," but "[t]he difficult task facing Darwinian evolution, however, would not be to remove parts from sophisticated preexisting systems; it would be to bring together components to make a new system in the first place." In that article, Professor Behe wrote that he hoped to "repair this defect in future work;" however, the court noted that he has failed to do so even four years after admitting his defect.[2, 21]

The *Kitzmiller* court also rejected the argument of irreducible complexity because it felt that it ignores another important alternative idea. The court stressed that although Professor Behe is adamant when he says a precursor "missing a part is by definition nonfunctional," that statement can legitimately be interpreted in a different way. Plaintiff experts pointed out it can also mean that a biological system will not function in the same way the system functions when all the parts are present. For example, in the case of the bacterial flagellum, removal of a part may prevent it from acting as a rotary motor moving the bacteria through its environment. However, the possibility that a precursor to the bacterial flagellum functioned not as a rotary motor, but in some other way, for example as a secretory system is extremely possible and even likely. This is a principle of evolution know as exaptation. Exaptation is widely accepted evolutionary concept that explains the evolution of complicated, multipart systems evolved through natural means. Exaptation proposes that some precursor of the subject system had a different, selectable function before experiencing the change or addition that resulted in the subject system with its present function. There were other examples of this phenomenon given at trial; for example, the evolution of the mammalian middle-ear bones from what had been jawbones of reptiles and early mammal-like predecessors.[2]

The *Kitzmiller* court also quoted the NAS position on the concept of irreducible complexity. According to the NAS, "[S]tructures and processes that are claimed to be 'irreducibly' complex typically are not on closer inspection. For example, it is incorrect to assume that a complex structure or biochemical process can function only if all its components are present and functioning as we see them today. Complex biochemical systems can be built up from simpler systems through natural selection. Thus, the "history" of a protein can be traced through simpler organisms... The evolution of complex molecular systems can occur in several ways. Natural selection can bring together parts of a system for one function at one time and then, at a later time, recombine those parts with other systems of components to produce a system that has a different function. Genes can be duplicated, altered, and then amplified through natural selection. The complex biochemical cascade resulting in blood clotting has been explained in this fashion."[2]

Next in its analysis of the scientific status of ID, the *Kitzmiller* court turned its attention to another foundation concept of ID. Defense experts used the phrase "purposeful arrangement of parts" to express the concept. Behe and defense expert Minnich believed that design is inferred when parts of a living system appear to be arranged for a purpose. Bebe continued that the strength of the inference is quantitative; the more parts that are arranged, the more intricately they interact, and the stronger the evidence for design. Bebe believed that the appearance of design in biology is overwhelming, with the best, most-rational explanation being an intelligent cause, although they refused to identify the designer.[2]

The *Kitzmiller* court rejected this analysis, stating that this argument is merely a restatement of nineteenth-century theologian Reverend William Paley's argument, except Paley had no hesitation in identifying the designer as the Judeo-Christian God. Paley believed that complex organisms must be designed. Plaintiff expert testimony asserted that this argument is not scientific inasmuch as it can never be falsified. This nonfalsification criticism was admitted by Behe.

The court concluded that this alleged positive argument for ID does not satisfy the ground rules of science, which require testable hypotheses based upon natural explanations; ID is reliant upon forces acting outside of the natural world, forces that we cannot see, replicate, control, or test, which have produced changes in this world.[2]

Finally, the *Kitzmiller* court dismissed ID as legitimate science by noting the complete absence of peer-reviewed publications supporting the theory. The court stated that expert testimony revealed that the peer-review process is "exquisitely important" in the scientific process. It is a way for scientists to write up their empirical research and to share the work with fellow experts in the field, opening up the hypotheses to study, testing, and criticism. Even defense expert Behe recognized the importance of the peer-review process. The court stated that evidence presented in this case demonstrates that ID is not supported by any peer-reviewed research, data, or publications. Once again, even Behe admitted that "There are no peer reviewed articles by anyone advocating for intelligent design supported by pertinent experiments or calculations which provide detailed rigorous accounts of how intelligent design of any biological system occurred." Additionally, Professor Behe conceded that there are no peer-reviewed papers supporting his claims that complex molecular systems, like the bacterial flagellum, the blood-clotting cascade, and the immune system, were intelligently designed.[2]

Next, and totally without necessity to reach a legal conclusion, the *Kitzmiller* court addressed ID's claims against evolution. The court stated that ID proponents support their assertion that evolutionary theory cannot account for life's complexity by pointing to real gaps in scientific knowledge. However, the court noted that an overwhelming number of scientists and scientific associations have rejected ID's challenge to evolution. For example, the court cited the plaintiffs' expert in biology, Dr. Miller, a professor at Brown University, who has written university-level and high-school biology textbooks used throughout the nation. Brown's testimony was unrebutted by the defendants. He stated that evolution, including common descent and natural selection, is "overwhelmingly accepted" by the scientific community and that every major scientific association agrees.[2]

The court continued that, despite the scientific community's overwhelming support for evolution, the defendants and ID proponents insist that evolution is unsupported by empirical evidence. The plaintiffs' science experts, Miller and Padian, explained how ID proponents generally, and *Pandas,* the book the board referred Dover students to and who all pertinent trial experts agree is representative of ID, specifically, distort and misrepresent scientific knowledge in making their antievolution argument.[2]

The court stated that based upon unrebutted plaintiff expert Padian's testimony, *Pandas* distorts key issues. For example, Padian testified that *Pandas* misrepresents classification theory important to evolution, the concept of homology that allows scientists to evaluate comparable parts among organisms for classification purposes, the concept of exaptation, and evidence for evolution in the fossil record.[2]

The court also cited plaintiff expert Miller, who testified that *Pandas'* treatment of comparative biochemistry is "inaccurate and downright false" and explained how *Pandas* misrepresents basic molecular biology concepts to advance design theory. In addition, Miller refuted *Pandas'* claim that evolution cannot account for new genetic information and pointed to more than three dozen peer-reviewed scientific publications showing the origin of new genetic information by evolutionary processes. Miller testified that *Pandas* misrepresents molecular biology and genetics principles, as well as the current state of scientific knowledge in those areas in order to teach readers that common descent and natural selection are not scientifically sound.[2]

The *Kitzmiller* court after these scathing attacks on ID, concluded that the proper application of both the endorsement test and *Lemon* test to the facts of the case makes it abundantly clear that the Board's ID Policy violates the Establishment Clause. The court continued that ID is not science and that it cannot be uncoupled from creationism, which previous court decisions—including the two from the Supreme Court of the United States, *Epperson* and *Edwards*—held is a religion.[2]

Concomitantly, the *Kitzmiller* court dismissed ID proponents' claims that evolutionary theory is antithetical to a belief in the existence of a supreme being and to religion in general, and that it is not good science. The court cited plaintiffs' experts who testified that evolution does represent good science, is overwhelmingly accepted by the scientific community, and that it in no way conflicts with, nor does it deny, the existence of a divine creator. In fact, the court stated that many evolutionists hold religious beliefs, including the belief in a creator.[2]

The court permanently enjoined the Dover School Board and School District from maintaining the ID Policy as a violation of the First Amendment to the United States constitution and to the equivalent provisions of the Pennsylvania constitution.[2]

The *Kitzmiller* decision was not appealed. All but one of the school-board members who endorsed the ID Policy failed to be reelected in November 2005, a month before the court decision. Their replacements did not support the teaching of intelligent design and did not appeal the decision[16] Therefore, the matter never found its way to a Federal Appeals Court or the Supreme Court of the United States.

As I reflect on the *Kitzmiller* case, it is historically clear that the modern ID movement began about the same time that *Edwards* was decided.[17] It was part of a creationist response to court decisions holding that scientific creationism is religion, most notably the McLean and Edwards decisions. As religion, creationism was banned from the public-school science curriculum. However, it should be noted that neither *Edwards* nor *Epperson*, the two evolution in public school Supreme Court cases, prohibits the teaching of creationism in other contexts, such as part of a literature or world religions class.[1, 9] The Supreme Court has also made clear in a number of cases involving the role of religion in schools that "the Bible may constitutionally be used in an appropriate study of history, civilization, ethics, comparative religion or the like, but not in science classes."[18]

Both *McLean* and *Edwards* addressed state legislation requiring the teaching of creation science when evolution is taught. *Edwards* was a more important decision since it is a Supreme Court case with national applicability; however, it dealt with a narrower range of issues than *McLean*. This is a general characteristic of Supreme Court cases. They tend to be narrow rulings, providing the minimum justification for the decision. In addition, *Edwards* was decided on summary judgment, without a trial. *McLean*, on the other hand, was decided after a full trial, with discovery, depositions, witness statements, testimony, cross examination, etc. The court record for *McLean* is consequently much deeper and broader than *Edwards*.[1, 4]

Both *Edwards* and *McLean* found creation science to be religion and therefore unconstitutional to include in the public-school science curriculum. *McLean* went further, using expert scientific testimony at a full-blown trial to address this question. The *McLean* decision presented a broad-based attack on the scientific status of creation science.[1, 4, 16, 19]

There are a number of similarities between *McLean* and *Kitzmiller* cases. Both were many-day trials. *McLean* dealt with whether the claimed alternative to evolution, creation science, was truly scientific; *Kitzmiller* dealt with whether intelligent design was truly scientific. The main legal concern of both cases was Establishment Clause violation, and both cases used the *Lemon* test to determine this (Kitzmiller also used the endorsement test, which was developed by the Supreme Court after *McLean* and *Edwards* were decided).[2, 4, 19]

A technical legal concern runs through the *McLean, Edwards,* and *Kitzmiller* cases. This concern was also articulated by Eugenie Scott of the National Center for Science Education, an organization that was part of the plaintiffs' legal team in *Kitzmiller*. Scott argued that the ID Policy was struck down by the court because ID is religious. However, she stated, that the *Kitzmiller* court, as did *McLean* before it, believed that bad science is not unconstitutional to teach; the First Amendment doesn't prohibit bad science, but the establishment of religion. Scott continued that although science was a secondary consideration in both *McLean* and *Kitzmiller,* she believed that it was necessary to the plaintiffs' success in both trials to address what it is and what it is not science.[1, 2, 4, 19]

As was articulated in the *Edwards* dissent by Justice Scalia, and also addressed by Scott, even though the establishment clause bans governmental advancement of religion, there are certain circumstances that allow limited governmental support of religion. If a law has a primarily secular (nonreligious) purpose and effect, it can be constitutional, even if it has a secondary benefit to religion. For example, laws that require a school district to purchase nonreligious books for parochial students can be legal even if religion benefits; the primary benefit of the law is improved education for students in the community. There is a secular reason for buying books for parochial students; therefore, the practice is constitutional. Extrapolating from this, Scott wrote that when you have a phrase like creation science, it is not jumping to conclusions to suspect that religion is central to the concept. Or with intelligent design, it is logical to ask, who is the designer? Once you have a creative agent, you have creationism, which is a religious concept, and hence unconstitutional to promote in public schools.[1, 19] However, as articulated by Scott, if a legitimate secular reason for teaching creationism could be devised, or if creationism were legitimate science, then the religious component could be considered secondary to the secular, and the law might pass constitutional muster.[19]

Defendants in *McLean, Edwards,* and *Kitzmiller* attempted to meet one or both of these exceptions. Defendants claimed that students would benefit educationally from being taught alternatives to evolution. *Kitzmiller* defendants even added that the ID Policy promotes critical thinking, and *Edwards'* defendants said it promotes academic freedom. Scott believed, as did the plaintiffs and judges in *McLean* and *Kitzmiller*, that plaintiffs had to counter these claims. Therefore, whether creation science or ID qualify as science became highly relevant to them in regard to showing a secular purpose for their teaching. There cannot be a legitimate secular purpose for teaching unscientific topics in a science class.[1, 2, 4]

The *McLean* plaintiffs' attempt to address this issue has been discussed in detail previously in this work. In *Kitzmiller,* plaintiffs' witnesses charged with addressing the definition of science gave similar, but slightly less expansive, testimony to that in *McLean*. They stressed that scientific explanations must be testable, restricted to natural causes, tentative, and falsifiable. Particularly, they focused on the point that scientific explanations are restricted to natural causes, or methodological naturalism as it has been called.[2, 4, 19]

The testimony caused the *Kitzmiller* court to fall into the same intellectual and ethical dilemma that befell *McLean*. It also purposely or negligently simplified the intellectual problems associated with the demarcation of science from nonscience. The *Kitzmiller* court, as *McLean* before it, did not have the academic background, credentials, or proper forum to enter into an intellectually complicated and specialized area such as demarcation, nor should it have done so.[2, 4, 19]

However, I believe that *Kitzmiller* did have the capacity to analyze the ID Policy's constitutional viability, using the *Lemon* and endorsement test, and find it wanting. Clearly, the ID Policy had a primarily religious purpose and effect, and it clearly entangled government in religious issues. That should have been the end of the matter, rather than engage in a judicial overreach that brings into question the court's integrity and ultimate wisdom. Once again, we see a proper decision, but made for wrong reasons, raising unsettling questions concerning the integrity of the courts and their decision-making processes.[2] Simplifying complicated demarcation issues for political reasons or facilitating desired judicial outcomes is not principled or acceptable.

The story of the teaching of evolution in public schools from *Scopes* through *Kitzmller* teaches much about science, law, and education. The ultimate outcome of these cases was, and is, important for many reasons.

Evolution is critical to biology. Not only is it an important biological concept, but also modern biology's main unifying principle. In addition, evolution is important to many other modern intellectual disciplines. Evolutionary and Darwinian ideas weave their way through physics, chemistry, history, sociology psychology, anthropology, and theology.

Teaching modern biology without evolution is intellectually unthinkable; teaching practically any intellectual discipline without evolution is also probably equally unthinkable. It must be included in modern biology education; it must be included in modern education. Remember the famous quote by Dobzhansky, "Nothing in biology makes sense except in the light of evolution." Biology without evolution is a collection of disparate facts about the living world; with evolution, biology is a coherent modern science. Evolution is so important to biology that many mark the beginning of modern biology to the publication of *On the Origin of Species* by Darwin in November of 1859.

In spite of its lofty status in biology and education, we have seen that evolution found its way several times into the United States courts during the twentieth century and even into the twenty-first century: the only modern intellectual idea to do so. This was for the most part, due to religious fundamentalist challenges to it being taught in public schools.

Litigation involving evolution allows us to explore some important questions. Among them are, does the evolution in public-school cases teach us anything about the nature of how United

States courts make decisions, what is the nature of science and the scientific method, and is science merely just an objective pursuit of truth? Secondary questions raised are what is so offensive and dangerous about evolution to creationists that its teaching in public school must be limited or ended? And at least since *Epperson*, I guess the converse question can be asked, what is so offensive and dangerous about creationism to many mainstream scientists and mainstream religious that it must be barred from science classrooms? Is no accommodation possible on either side?

In this work, I sought answers to these questions, analyzing the evolution in public-school cases within the context of our changing views of evolution and how it works. These cases span more than eighty years and include two United States Supreme Court decisions.

As has been discussed, the *Scopes* trial, which marks the beginning of the evolution cases in the United States, occurred in the 1920s. Although almost no scientist questioned the fact of evolution, intense debate raged over what mechanisms drive evolutionary change. Darwin's natural selection as a mechanism driving evolutionary change, so important to evolution today, was widely rejected by scientists at that time.

In *Scopes*, William Jennings Bryan attempted to use the scientific community's uncertainty over evolutionary mechanisms to his advantage, questioning the value of a theory whose action couldn't be determined with some degree of certainty. Two additional important legal issues separate *Scopes* from subsequent evolution cases that we discussed in this work. First, *Scope* was decided exclusively in Tennessee state courts, and it never reached the federal court system. Second, it was decided before the First Amendment to the federal constitution was made applicable to the individual states. Remember, before *Cantwell v. Connecticut* and *Everson v. Board of Education* cases in the 1940s, the religious clauses of the First Amendment of the United States Constitution only applied to the federal government; they were not binding on the states. This left only the Tennessee constitution as a way of invalidating the Butler Act litigated in *Scopes*.

Looking back from a twenty-first century perspective, the *Scopes* court's decision holding that the Butler Act did not violate the

Tennessee constitution rested on the disingenuous conclusion that since members of various Christian churches differed in their views on evolution, the Tennessee state constitution's prohibition that the state must not show preference to religion or mode of worship was not violated. No significance was given by the *Scopes* court to the act's obvious preference and protection of a particular interpretation (Fundamentalist) of a Jewish/Christian Holy Book, or to that Holy Book itself. The court also rejected that the Tennessee legislature's motive in passing the act had any legal significance. A legal analysis ignoring the motives of the legislature in passing an act has been consistently rejected in subsequent federal courts.

The *Scopes* court's actions left only vagueness as the only possible constitutional challenge to the Butler Act under Tennessee law. However, even this challenge was precluded when the Tennessee state court removed vagueness as an issue by cynically interpreting the statute as only prohibiting the teaching that humans evolved, rather than imputing broader meaning to it.

Without a path to make a state or federal constitutional challenge, the *Scopes* case died on a technicality. The narrow interpretation of the act was designed to preclude any need by the court to address the scientific legitimacy of evolution. The court and prosecution knew it wouldn't be able to do so because of the almost unanimous support among scientists for evolution even at that time. The end result was that the Tennessee Supreme Court concluded that the Butler Act was merely a valid exercise of a state's right to regulate the activity of employees while performing their job

The *Scopes* case never reached the federal court system or the United States Supreme Court as the defense had hoped. However, what we see in almost all subsequent evolution cases is similar cynical and disingenuous legal maneuvering on one side or the other that characterized *Scopes.*

One additional point before moving on, least one thinks that somehow only science and religion were the only issues at stake in *Scopes,* or that somehow those that opposed the Butler Act were on the meritorious side of things in a modern interpretation of that view, please consider this. As we have discussed in Chapter 1, there was at

least an indirect link between evolution and a host of abhorrent political and social movements of that time. Bryan saw this, but he wasn't
able to separate the science of evolution from the politics and a culture that coopted evolutionary ideas for unsavory social, economic,
and political agendas. For example, Bryan is often portrayed as a religious zealot because of his opposition to evolution in public schools.
Bryan has also often been criticized by historians of science for falsely
linking evolutionary theory, particularly Darwin's theory of natural selection, to certain political and social movements that Bryan
found, and our society today largely finds, abhorrent. However, his
opposition was not without some justification. Don't overlook that
the very book used by John Scopes, *Hunter's Civic Biology*, was blatantly racist, representing the Caucasian race as "finally, the highest
type of all" and proposing involuntary sterilization as a remedy for
crime and immorality. Bryan's motivation in opposing evolution in
public schools was not solely religious. Although he understood that
a link existed between evolution and these social movements, Bryan
failed to recognize that evolution itself was being co-opted to support political and social agendas that preceded Darwinian evolution,
and that separate from evolution the unsavory social movements
were structurally embedded in Western culture. Therefore, any links
between evolution and these movements is much more complicated
and nuanced than Bryan imagined.

In addition, also as discussed in Chapter 1, one must remember
that Bryan never supported the teaching of *Genesis* or creationism in
public schools. He strongly believed that neither religion nor evolution were morally or constitutionally proper subject for inclusion
into the curriculum by legislation.

As we discussed in Chapter 2, the trajectory of evolution in
public school cases changed permanently and drastically with the
Epperson v. Arkansas case, where an Arkansas law similar to the Butler
Act was decided in very different manner than what we saw in *Scopes*.
Recall that the Tennessee Supreme Court held in *Scopes* that the
Butler Act was merely a valid exercise of a state's right to regulate
the activity of employees while performing their job. It dismissed
claims that the act violated any prohibition regarding the establish-

ment of religion. In *Epperson*, we see a very different decision when the Supreme Court of the United States held that the Arkansas law violated the First Amendment of the United States Constitution. As such, any legitimate right a state might have to regulate employee activity must be subordinated.

The legal foundation for the *Epperson* decision was laid down in the 1940s when the United States Supreme Court in *Cantwell* and *Everson* made the two religious clauses of the First Amendment applicable to and binding on the individual states. As we discussed in Chapter 2, prior to these decisions, the First Amendment religious clauses only applied to the Federal government, not the states. As a result, the defense team in the *Scopes* trial did not have legal precedent necessary to effectively argue that the Butler Act violated those clauses (an issue paramount in *Epperson*). By the time *Epperson* was decided, however, it was well established constitutional law that the First Amendment applied to the individual states as well as the federal government. In addition, by 1963, a constitutional test for religious neutrality had been formulated by the Supreme Court of the United States in *Abington School District v. Schempp*, and it was made part of the analysis used by the *Epperson* court in determining whether the Arkansas law met First Amendment constitutional standards. This test was obviously not available in *Scopes*. Under the requirements of this test, the primary purpose or the primary effect of a state action must neither advance nor inhibit religion. Remember, the *Scopes* court had specifically rejected any purpose analysis stating that only the effect of a statute, rather than any "proclaimed motives" of the legislature or the statute itself are determinative of an act's constitutional validity.

Besides significant changes in the legal landscape of First Amendment jurisprudence, in Chapter 2, we also discussed the significant scientific advances made in evolutionary biology between *Scopes* and *Epperson* decisions. I believe that the *Epperson* decision was influenced by this. Evolutionary biology as a scientific theory matured greatly during this time period and the *Epperson* decision implicitly, if not explicitly, recognized this. The emergence of the New Synthesis of Evolutionary Biology in the 1930s through the

1950s seemed to vindicate Darwin's natural selection as the prime mechanism for evolutionary change in the minds of most biologists. By the 1960s, widespread consensus on what drives evolutionary change prevailed in a post-New Synthesis scientific era that was not present at the time of *Scopes*. Remember, during the 1920s, although evolution was accepted almost unanimously by the scientific community, there was widespread disagreement as to what mechanisms drive evolution. After the Synthesis, evolutionary biologists generally recast evolution mostly in terms of Darwinian selection theory.

As discussed in Chapter 2, analysis of the *Epperson* decision illustrated that the court's confidence in evolutionary theory mirrored the confidence of the scientific community, and this was a determinative factor in the decision and in the other court decisions discussed in this work. Evolution was riding a wave of relative scientific consensus, which the court seemed to accept without reservation, analysis, or question. Yet the court failed to acknowledge this and instead claimed that its decision to hold the Arkansas law unconstitutional was based entirely on the supposed sole religious purpose or motivation of the Arkansas legislature in enacting it. A critical look at the facts in Chapter 2 of this work showed this position to be disingenuous at best. Rather, the decision was crafted as a result of the court's belief in the scientific status of evolutionary theory and an unwillingness to fairly consider a challenge to it.

As seen in Chapter 2, the *Epperson* decision specifically drew language verbatim from *amici curiae* briefs by the National Education Association of the United States and the National Science Teachers Association filed on behalf of the plaintiff. The court's almost complete reliance on these briefs supported its willingness to accept mainstream science's view on evolution. Further support for this assertion is seen in the court's statements that creationism is merely an anachronism that is indefensible in light of modern scientific knowledge. This may be true, but the court couldn't have fairly concluded this based on the analysis it articulated!

As was also discussed in Chapter 2 of this work, the court's acceptance of the scientific status of evolutionary theory provides an explanation for its apparent selective and biased use of the history of the law litigated in *Epperson*. As Larson stated, the court found it insignificant or ignored that two-third of the Arkansas voters, far more than can be attributed solely to fundamentalism, supported the law. Also, those that supported the law had mixed motives. For example, Bryan supported and campaigned for the Butler Act and the Arkansas law litigated in *Epperson* for many reasons; he was as concerned with preventing war and social exploitation as he was religion. Remember, Bryan was convinced that these social ills followed unfailingly from Darwinism. Further, as Ronald Numbers pointed out, studies have shown that the people who joined Bryan's antievolution crusade came from all walks of life and all parts of the country. In essence, the *Epperson* court provided little to support its conclusions that the sole purpose of the Arkansas law was religious.

Remember also the accusation made about Supreme Court Justice Fortas, who wrote the *Epperson* decision. He was accused, as we previously discussed, of altering the meaning of an advertisement the court used to support its sole purpose interpretation of the Arkansas law. The unaltered version supports a different stated purpose than the one articulated by the court. The unaltered version supports a purpose of promoting religious neutrality, rather than a promulgating Fundamentalist religion. Bryan, a proponent of the act, assumed that creationism could not be legally taught in public schools, and therefore he sought neutrality by also removing evolution.

I must reassert an important point made previously. Nothing in this discussion is meant to deny that religious motivation was behind the enactment of the Arkansas law, or to support any creationist claim against evolutionary theory's scientific legitimacy. Evidence for religious motivation is found from the express language of the law, as well as significant Fundamentalist religious involvement in its enactment. Notwithstanding this, the *Epperson* decision is flawed. The court's purpose analysis is tainted by its own notion of what constitutes scientific legitimacy and by its own bias that evolutionary theory should stand alone in the public-school science curriculum. In *Epperson*, it appears that the Supreme Court accepted the view of the mainline scientific community, without critical analysis, and then contrived to use the *Schempp* purpose test in a contorted and perhaps dishonest manner, to support its preconceived notion. One wonders just how pervasive this flawed approach to judicial determination is in federal and state courts in other cases and in other matters. In *Epperson*, it's not so much what the court decided. I agree that the statute in question violated the First Amendment; it's rather how the decision was reached that I call into question. *Epperson* would have been on a much stronger legal and ethical foundation to rule that the predominate purpose and effect of the Arkansas Statute was religious in nature, which is sufficient to invalidate it, and to give judicial notice that under no reasonable or generally understood definition can evolution be defined as religious.

As we have seen, however, *Epperson* did not end legal controversy concerning evolution in public schools; it only marked the beginning of federal court involvement. Since *Epperson*, all antievolution statutes or state actions designed to limit the teaching of evolution in the schools have been cast as First Amendment religious-clause issues.

In Chapter 3, we discussed how after *Epperson* creationists realized that it would be constitutionally fatal to attempt to eliminate evolution from the public-school science curriculum with a *Scopes*-type statute. Therefore, they switched course and attempted to achieve their political and religious objectives by demanding that creationism be given equal time with evolution in the public-school science curriculum. The first case that dealt with this issue was *Wright v. Houston Independent School District*. Notwithstanding this change in direction by creationists, the *Wright* court rejected plaintiff/creationist assertions that evolutionary theory is a religion as that term is constitutionally understood. It also rejected a requirement that creationism must be given equal time in the science curriculum of public schools with evolution.

The *Wright* decision showed some of the same analytical problems as *Epperson;* it presumed that evolutionary theory is a scientific theory, which should stand alone in the public-school science curriculum. Once again, we see a federal court adopt mainstream science's view on this matter without any real analysis or explanation. However, there are analytical differences between *Wright* and *Epperson* that make *Wright* much easier to justify. No strained analy-

sis was done by the *Wright* court as was done in *Epperson*. A creationist religious motive and effect to brand evolution a religion and to
mandate equal time for creationism was palpable. Courts have a right
to take judicial notice of the obvious, and two things were obvious
in *Wright*: evolution is not religion by any reasonable and generally
understood definition of that term, and the motive and effect of the
plaintiff's actions were obviously religious. Judges do not live in a
vacuum but within a society. *Epperson* would have been well advised
to take the more constrained position similar to *Wright* in its analysis.

Daniel v. Waters represents a case where a Tennessee equal-time
statute was tested in a federal court. This case is important for two
reasons. It marks the first time a federal court was presented with an
equal-time statute, and it was the first time the *Lemon* test was used
to determine the constitutionality of a statute in an evolution case.
The statute at issue in *Daniel* was blatantly religious in nature. The
Daniel court responded to it by invalidating the statute as violating the Establishment Clause of the First Amendment to the federal
constitution. On analysis by the court, the statute failed to meet the
requirements of all three parts of the *Lemon* test. Likely, because of
the obvious and blatant religious nature of the statute in question,
the *Daniel* court did not feel a need to address the scientific status
of evolutionary theory as *Epperson* and *Wright* did; rather, it focused
solely on the statute in question. *Daniel* is a sound, judicially constrained decision, legally grounded in the law of its time. It did not
contain the judicial overreach that will be seen in subsequent equal
time evolution cases, especially *McLean* and *Kitzmiller*. In many
ways, *Daniel* is a model for cases of its type.

Daniel also did not end equal-time legislation. More refined
and sophisticated equal-time statutes, in the form of balanced
treatment, followed the *Daniel* decision. These statutes cynically
attempted to circumvent the constitutional challenges posed by the
Epperson Supreme Court case. Two statutes of this nature, enacted in
the 1980s, had a profound effect on evolution in public-school litigation. One statute was enacted in Arkansas, the other in Louisiana.
The cases involving these statutes are controversial. The first is
McLean v. Arkansas; the second is *Edwards v. Aguillard. Edwards*, as

did *Epperson* before it, also found its way to the Supreme Court of the United States.

McLean v. Arkansas has been hailed by *Science*, a leading journal of American science, as "the finest legal document ever written about this question—far surpassing anything that the *Scopes* Trial generated, or any document arising from the two Supreme Court cases... Judge Overton's (the federal court judge who decided *McLean*) definitions of science are so cogent and clearly expressed that we can use his words as a model for our own proceedings." As cited in Chapter 4, *Science* published Judge Overton's opinion verbatim as a major article. Notwithstanding this, others have interpreted *McLean* very differently. As we have discussed in Chapter 4, it has been criticized as representing an overbroad opinion that overestimates the open-mindedness of scientists and ignores the dangers inherent in a federal court strictly defining subjects within the public-school curricula, especially subjects that it has no expertise on or legitimate ability to address. I tend to agree with the latter interpretation.

A critical review of the *McLean* decision reveals important intellectual and legal issues and problems. As was discussed in Chapter 4, the *McLean* court disregarded certain legal precedent of that time period in reaching its decision. Although few would disagree that the history and legislative history of the Arkansas statute established a clear religious motive or purpose, the *McLean* court ignored a key constitutional question: was religion the sole purpose of the act? This was not a trivial question at that time, although it was subsequently addressed in the *Edwards* Supreme Court case. *McLean* was decided in 1982. Up to that time, the Supreme Court of the United States had invalidated a state statute for lack of secular purpose (what has become the first part of the *Lemon* test) only twice before; in *Epperson,* the evolution case in 1968, and in one other case, *Stone v. Graham* in 1980. In *Stone,* the Supreme Court ruled that a Kentucky law that required the posting of the Ten Commandments on the wall of every public school classroom in the state violated the Establishment Clause of the First Amendment because the purpose of the display was religious. In both cases, the Supreme Court found that the respective statutes were solely or wholly motivated by religious purpose, a weighty standard of review. However, despite this precedent, the *McLean* court engaged in no meaningful analysis of whether a finding of sole religious purpose was necessary to invalidate any statute, including the Arkansas Act, for violating the first prong of the *Lemon* test. Rather, the court merely concluded that the correct standard of review should be a specific purpose or primary purpose test, with no discussion of the relationship of this test to a sole purpose test. Once again, we see court that seems intent on reaching a particular verdict, whatever it had to ignore or overlook to do so.

A second analytic problem associated with the *McLean* decision had to do with the court's conclusions on the motive of the Arkansas legislature in enacting the act. As discussed in chapter 4, the court concluded that the expressed motives of the draftsperson and sponsors of the act (some of whom were not even from Arkansas) were sufficient to support the conclusion that the primary or specific purpose of the Arkansas legislature in passing the act was religious. However, this view of the court causes a legitimate question to be raised. How could a court impute universal motives on a group from the actions and comments of a few members, even when those few have a leadership capacity, especially since some weren't even from Arkansas? This is not to say that the statute in question should have survived a *Lemon* analysis. Even the act's definition of creation science betrayed its religious purpose as thinly veiled fundamentalism. Once again, it's how the decision was rendered that is problematic to me. Procedural fairness must matter. It's the basis of being a nation governed by laws, which the United States purports to be.

The court did not need to continue its analysis after the statute failed a *Lemon* purpose analysis. A statute must pass all three prongs of the test to survive Establishment Clause scrutiny. Failure to pass even one prong is constitutionally fatal. Notwithstanding this, *McLean* showed an absence of judicial restraint by entering into a detailed, but misleading, discussion of what science is, and what it is not.

After hearing plaintiffs' experts, *McLean* concluded that the essential characteristics of science include naturalness, tentativeness, testability, and falsifiability. The court stated that creation science failed to meet these criteria; therefore, it is not science but religion. However, in doing so, as we discussed in Chapter 4, the *McLean* court and plaintiffs' witnesses simplified a complicated demarcation of science/nonscience argument in a way that was disingenuous at best and deliberately misleading at worst. The court and plaintiffs' witnesses also selectively ignored or downplayed the significance of philosophers who basically rejected the court's definition of science and how it works.

This raises an important issue in a whole line of evolution in public-school cases, from *Scopes* to *Epperson*, through *McLean*, *Edwards,* and *Kitzmiller.* That is, the less-than-honest arguments of participants on both sides designed to support their particular position. Once again, one wonders if these cases act as a model illustrating a pervasive and structural problem associated with the judicial system in this country.

As we discussed in Chapter 4, there is little doubt that creation scientists misrepresented the religious nature of their discipline. Creation-science literature is replete with religious references. For example, as *McLean* articulated, Duane Gish, in *Evolution: The Fossils Say No!* stated, "By creation we mean the bringing into being by a supernatural Creator of the basic kinds of plants and animals by the process of sudden, or fiat, creation. We do not know how the Creator, created, what processes He used, for He used processes which are not now operating anywhere in the natural universe... We cannot discover by scientific investigation anything about the creative processes used by the Creator." Also, Henry Morris, called by some the founder of the creation-science movement, said, "it is... quite impossible to determine anything about Creation through a study of present processes, because present processes are not created in character. If man wishes to know anything about Creation (the time of Creation, the duration of Creation, the order of Creation, the methods of Creation, or anything else) his sole source of information is that of divine revelation...we are completely limited to what God has seen fit to tell us..." And a chief creation-science organization, the Creation Research Society, must sign specifying their beliefs,

"¹The Bible is the written Word of God, and because we believe it to be inspired throughout, all of its assertions are historically and scientifically true in all the original autographs. To the student of nature, this means that the account of origins in *Genesis* is a factual presentation of simple historical truths.⁴ Finally, we are an organization of Christian men of science, who accept Jesus Christ as our Lord and Savior. The account of the special creation of Adam and Eve as one man and one woman, and their subsequent fall into sin, is the basis for our belief in the necessity of a Savior for all mankind. Therefore, salvation can come only thru accepting Jesus Christ as our Saviour."

Yet despite all the evidence that creationism has a strong religious component and overlay, creationists audaciously claimed that there is no reason why their model of origins should not be accorded equal space with evolution within the scientific curriculum, and that creationism is an unbiased search for truth about how species, including humans, got here.

However, does the religious nature of creationism provide the sole answer to the question of why the mainstream scientific and legal communities have such deep-rooted animosity toward creation science? The *McLean* court would have us believe that it does, and that the court is motivated only by a desire to protect our youth from unscientific ideas posing as legitimate science. I believe that this certainly is a motive. However, as discussed in Chapter 4, Ronald Numbers suggests a possible additional reason for the animus toward creationism. Numbers, relating the opinion of certain sociologists of science, contended that the *McLean* trial provided a revealing glimpse of scientists vigilantly guarding their boundaries. Numbers believed that just as creationists pushed the limits of science to accommodate their religiously inspired agenda, so their opponents invoked a narrow definition of science to maintain their "monopoly over the market for 'scientific' knowledge in Arkansas schoolrooms." Numbers suggested that by discrediting the creationists as "pseudoscientists" unworthy of public patronage, the established scientific community hoped to eliminate a politically powerful competitor for scarce resources. Numbers stated that whatever the respective merits of the two sides, their struggle illustrates the historically contingent nature of "science" and the futility of assigning to the term an invariant meaning.

The *McLean* trial, according to Numbers, also shattered myths about the so-called warfare between science and religion. The *McLean* plaintiffs were an eclectic group: a scientific society, a teacher, a science teachers' organization, various clergy from a variety of Christian denominations, and various Jewish organizations. Yet no religious group appeared on the list of defendants supporting creationism. Further, testifying at the trial against creationism were Christian clergypersons. In contrast, most of the witnesses in support of the act were well-credentialed scientists. Numbers related that, given the composition of the two sides, Langdon Gilkey, a theologian and participant, characterized the controversy as involving "two bizarre, unaccustomed and visibly uneasy partnerships: on the one side a union of what we might call elite religion and elite science, and on the other side a union of 'popular' (fundamentalist) religion with 'popular science.'"

The *McLean* decision's obvious intent was to establish the scientific legitimacy of evolutionary theory, while demonstrating that creation science failed to meet any reasonable criteria for inclusion as a scientific theory. In doing so, as has been discussed, the *McLean* court went far beyond any previous Federal court decision, or what was prudent or even defensible.

Whatever its failings, the *McLean* court held that creationism was merely a pretext for the propagation of Fundamentalist religious beliefs, and the Arkansas Statute was held to be unconstitutional as violating the Establishment Clause of the First Amendment. The *McLean* decision unambiguously established that balanced treatment statutes faced a tough Establishment Clause challenge in the federal courts. Notwithstanding this, balanced treatment legislation and litigation did not end. In Louisiana, a related statute was subsequently enacted by the state legislature and was quickly challenged in the federal courts, ultimately in the Supreme Court of the United States in *Edwards v. Aguillard.*

As we have seen, both the *McLean* and *Edwards* cases occurred within a time period characterized by increased creationist lobbying in the state legislatures for equal time legislation. From 1980 to 1985, such bills were introduced into nineteen state legislatures. Arkansas and Louisiana were two states that enacted the legislation into law. The Louisiana Creationism Act involved the Supreme Court of the United States in a difficult constitutional question. The issue framed by the court in *Edwards* was a narrow one, which is common in Supreme Court litigation; did the act violate the Establishment Clause of the First Amendment in the special context of the public-school system?

As discussed in Chapter 5, the court historically has given public schools considerable discretion in developing curricula as they see fit. However, despite this, the court has been particularly vigilant in monitoring compliance with the Establishment Clause, often invalidating various types of statutes, which advance religion in public schools. After all, this makes a lot of sense; kids are impressionable and a captive audience.

The test for Establishment Clause validity at the time *Edwards* was decided was still the *Lemon* test. Under *Lemon*, the purpose of a legislature in enacting an act into law receives particular consideration. This is a complex area of the law as evidenced by the different legal positions the Supreme Court justices took on this matter in *Edwards*. Although a majority of seven of nine justices found that the statute violated the First Amendment for lack of secular purpose, the decision was contentious for many reasons.

As we have seen in Chapter 5, Justice Scalia critiqued the majority's holding that the Louisiana Act was unconstitutional because its primary purpose was to endorse a particular religious doctrine. In support of his critique, he cited precedent from certain prior Establishment Clause Supreme Court cases. Scalia maintained that such precedent requires a statute to be wholly or entirely religious in purpose before it can be invalidated. He maintained that since the Creationist Act's stated purpose was a secular purpose of protecting academic freedom, and in his opinion the legislative history of the act seemed to provide at least limited evidence of this, the statute should not have been invalidated by summary judgment.

In sharp disagreement with Scalia, after reviewing the act's legislative history, the majority of justices in *Edwards* rejected the Louisiana legislature's stated purpose of protecting academic freedom as a sham, maintaining that the statute did not and could not serve this purpose. Rather, the majority believed that its purpose was to discredit evolution. The majority asserted that the act served to restructure the public-school science curriculum, selecting out for special treatment one scientific theory, evolution, which has historically been disfavored by certain religious sects. However, Scalia's position was nuanced. He was not asserting that the statute was constitutionally problem-free (although that was certainly the implication from his analysis); rather, he was stating that the correct way to determine the question of its constitutionality was an evidentiary trial, not summary judgment.

The majority, rejecting any scientific validity for creation science, concluded that creation science embraced a religious doctrine and included a belief in the existence of a supernatural creator. The majority believed that the act's preeminent purpose was to advance a religious point of view or to prohibit evolution from being taught. The *Edwards* court majority believed that creation science is not legitimate science and that the secular purpose necessary to pass constitutional musters under Establishment Clause criteria was nonexistent. The majority's opinion relied in large part on the historical and contemporary link between creation science and fundamentalism and the reliance of creation science on supernaturalism; that is, its dependence on a creator.

The Louisiana statute did not define creation science as did the Arkansas Statute litigated in *McLean*. However, the Louisiana Statute litigated in *Edwards* did set forth a two-model approach in similar fashion to the Arkansas Statute. This two-model approach, historically espoused by creationist literature, asserts that there are only two possible explanations for earth's life history—creationism or evolution. Under this approach, any evidence against one model functionally serves as evidence in support of the other. Historically, the evolutionary component has been defined by creationist literature in traditional Darwinian mechanistic terms; that it, a reliance on natural selection as the major driving force in evolutionary change. Both the *McLean* court and *Edwards* majority constitutionally rejected the two-model approach as motivated by religious purpose. In doing so, they accepted the evidence presented by the mainstream scientific community that the two-model system is logically and functionally flawed.

It must be admitted that to accept a two-model approach one has to ignore the history of creationism and creation science in the United States and the attempts of creationists to insert creationism into the United States science curriculum. Further, one has to ignore the many alternatives to Darwinism and creationism that have historically been espoused by various scientists and religions. Since the 1970s, many scientists were particularly active in the development of alternative evolutionary theories and mechanisms. These alternatives have generated extraordinary and intense debate within the scientific community, and clearly some are pseudoscience. As discussed previously, alternatives include punctuated equilibrium, pangenesis, Gaia theory, saltation theory, neutral theory, and others. However, none of these alternative theories reject naturalistic mechanisms or interpretations or espouse supernatural ones. In addition, there are even several alternative supernatural theories to creation and evolution, some ancient, including many associated with Eastern religions, other religious traditions, and philosophy. The two-model approach is just not sustainable when scrutinized objectively, and Scalia should have known better than to try to legitimate it.

Mainstream scientific testimony further convinced the *Edwards* court majority that no real scientific evidence could legitimate a creationism model because of its ultimate reliance on nonnaturalistic— that is, supernatural—explanations. The majority accepted this position and expected that others of "honest" intent would do the same. Therefore, for purposes of the first prong of the *Lemon* test, the court majority held that the statute must fail for lack of secular purpose. In doing so, the court rejected any attempt to limit the teaching of evolution in the public-school system, either by banning it entirely, or by attaching the teaching of creationism or creation science as a condition to evolution being taught.

Once again, the majority decision in *Edwards* afforded the presumption of legitimacy that Federal courts have historically afforded mainstream scientists and philosophers espousing the scientific nature of evolutionary theory. They did this while disregarding the testimony of creationists and their expert witnesses as religiously motivated, even when the experts were well-credentialed. As much as it is clear from the legislative history of the act—and the history of creationist attempts at limiting the teaching of evolution in public schools through legislation—that the act was religiously motivated, it is equally clear that the Supreme Court in this case continued the bias shown by federal courts in accepting, *carte blanche,* views of mainstream science without serious question or even the slightest question. In doing so, they affirmed the summary judgment imposed by a lower court, thereby denying creationists an evidentiary trial.

In stating that I have, I do not suggest that if evolution was scrutinized by the court, it would have been found wanting as science. Evolution is a bed rock of modern biology. My main objection is a procedural one, rather than a substantive one. However, it must be admitted that evolution is different from most other areas of science because of its direct and palpable impact on so many other intellectual, political, and religious ideas. As a general rule, I don't believe the courts that addressed the evolution in public-school cases afforded this historical fact the sensitivity it perhaps deserved. This will be discussed in more detail a bit later.

As discussed in Chapter 5, problems exist in justifying the majority opinion in *Edwards.* Although nowhere near as broadly or explicitly as was done in *McLean,* the majority entered into the difficult philosophical terrain of defining which elements constitute legitimate science. Ironically, in doing so, they also fell into a two-model approach, albeit of different type than the creationists. The basic philosophical and ideological underpinnings of the majority caused it to view the controversy also in binary terms, evolutionism, which the court considered a type of naturalism, encompassing a materialistic view of the universe versus supernaturalism. The majority viewed creationism's ultimate reliance on the supernatural, and its undisputable historical Fundamentalist ties as containing fatal religious elements for Establishment Clause purposes. This may be a very legitimate and defensible constitutional approach and justify denying creation science a place in the public-school science curriculum. However, to take it a step further and enter the difficult terrain of defining science and demarcating science from nonscience creates the same type of problems that we saw in *McLean* and *Epperson*. Basically, the majority philosophically adopted the simplistic position that legitimate science can only be as the court defined it, and that became the most important basis of the decision. In effect, the majority believed that no reasonable scientist or person could consider creationism science because of its supernatural underpinning; therefore, the statute must have been motivated by default by religious purpose. Although the *Edwards* Supreme Court majority took a much more judicially restrained position on this issue than did the *McLean* district court, their basic conclusions in this regard were the same and demonstrated the same bias. It ignored, as did *McLean*, an entire body of work by philosophers of science on demarcation of science from nonscience as unimportant to the decision. The majority lacked a nuanced and a legitimate position in this regard that it should have recognized.

As has been discussed in Chapter 5, the *Edwards* majority and the *Edwards* dissent in arriving at their respective positions relied heavily on the materials presented to the court by the involved parties and interested individuals and organizations. A number of legal briefs were filed on behalf of both the appellants and appellees in the *Edwards* case in order to influence the justices. For the most part, the court majority adopted positions espoused in certain appellee briefs. All of these briefs maintained that the primary purpose of the act was to advance a particular religious belief and that the act singles out evolution among the sciences for prejudicial treatment. A number of these briefs stressed an argument of the type stated in *McLean*; that is, science has certain characteristics, such as it is guided by natural law, it has to be explanatory by reference to natural law, it is testable against the empirical world, its conclusions are tentative, and it is falsifiable. All of these briefs argued that creation science fails these stated criteria and therefore is simply not science; rather, it is an extension of Fundamentalist religious doctrine.

Also, we have cited that the appellant/creationist position was also supported by a number of briefs. The brief on behalf of the Catholic League argued that the Appeals Court panel opinion was erroneous because *Lemon* requires that a law must be entirely motivated by a purpose to advance religion, which the law in question was not. This basically was the position adopted by the Supreme Court dissent but rejected by the *Edwards* court majority.

The State of Louisiana Brief was especially detailed and took issue with the problem of demarcation of science and nonscience. The Louisiana Brief argued that each of the *McLean* requirements for what constitutes science is rejected by a majority of philosophers of science. Further, it maintained that among philosophers of science there is little agreement concerning criteria for the scientific character of theories. The State of Louisiana Brief maintained that in considering the various proposed definitions of science, no group of philosophers of science has formulated a definition that even remotely resembles the definition set forth in *McLean* except the ACLU expert witnesses. The Louisiana Brief asserted that the Supreme Court should not restrict the definition of science to this viewpoint. It further contended that there are a variety of definitions of science, with the *McLean* view a positivist and materialist minority approach, one that limits science to natural laws and insists on verification (testing and falsifiability) of all events in order to be scientific, a minority view. Despite this argument, the *Edwards* court still denied an evidentiary trial.

We have seen that those that opposed creationist attempts to eliminate or limit evolution's place in the science curriculum of public schools have been extremely successful in the federal courts. *Edwards* is an example of that success. However, it did not end Fundamentalist legal challenges to an evolutionary monopolized science curriculum. Therefore, this peculiar legal battle continued into the twenty-first century in a new incarnation of creation science, intelligent design (ID).

In Chapter 6, it was cited that the modern ID movement began about the same time that *Edwards* was decided. It was in part of a creationist response to court decisions holding that scientific creationism is religion and banning its teaching as part of the public-school science curriculum. This was the basis of the *Kitzmiller* case.

As Eugenie C. Scott stated (see Chapter 6), there are many similarities between the *Kitzmiller* and *McLean* cases. The *Kitzmiller* decision dealt the same fate to ID that *McLean* did to creation science. As with *McLean*, the *Kitzmiller* decision was not appealed, but stayed at the federal district court level. So the matter never found its way to a Federal Appeals Court or the Supreme Court of the United States. Both were many-day trials. *McLean* dealt with whether the claimed alternative to evolution, creation science, was truly scientific; *Kitzmiller* dealt with whether intelligent design was and truly scientific. The main legal concern of both cases was Establishment Clause violation, and both cases used the *Lemon* test to determine this (Kitzmiller also used the endorsement test, which was developed by the Supreme Court after *McLean* and *Edwards*). I might also add that, in my opinion, both *McLean* and *Kitzmiller* produced decisions that were overreaching and overbroad, and the defendants in both cases set forth an extremely weak defense of the statutes at issue.

A technical legal concern runs through the *McLean, Edwards,* and *Kitzmiller* cases. This concern was also articulated by Eugenie Scott of the National Center for Science Education, an organization that was part of the plaintiffs' legal team in *Kitzmiller*. Scott argued that the ID Policy was struck down by the court because ID is religious. However, she stated that the *Kitzmiller* court, as did *McLean* before it, believed that bad science is not unconstitutional to teach; the First Amendment does not prohibit bad science in her opinion, but the establishment of religion. Scott believed that although science was a secondary consideration in both *McLean* and *Kitzmiller,* it was necessary to the plaintiffs' success in both trials.

As was articulated in the *Edwards* dissent by Justice Scalia and also addressed by Scott, even though the Establishment Clause bans governmental advancement of religion, there are certain circumstances that allow limited governmental support of religion. If a law has a primarily secular (nonreligious) purpose and effect, it can be constitutional, even if it has a secondary benefit to religion. For example, laws that require a school district to purchase nonreligious books for parochial students can be legal even if religion benefits; the primary benefit of the law is improved education for students in the community. There is a secular reason for buying books for parochial students; therefore, the practice is constitutional. But this led to a real concern that runs through *McLean, Edwards,* and *Kitzmiller.* As stated by Scott, if a legitimate secular reason for teaching creationism could be devised, or if creationism was legitimate science, then the religious component could be considered secondary to the secular, and the law might pass constitutional muster.

Defendants in both *McLean* and *Kitzmiller* attempted to meet one or both of these criteria in defense of the law or policy in question. Defendants in both cases claimed that students would benefit educationally from being taught alternatives to evolution. *Kitzmiller* even added that the ID Policy promotes critical thinking. Scott believed, as did the plaintiffs in *McLean* and *Kitzmiller*, that plaintiffs had to counter these claims. Therefore, whether creation science or ID qualify as science or weak science became relevant to them. There cannot be a legitimate secular purpose for teaching unscientific topics in a science class.

Unfortunately, the *McLean* and *Kitzmiller* plaintiffs believed that to combat this they had to show that evolution was valid science and creation science (*McLean*) and ID (*Kitzmiller*) were not. Hence, they entered the difficult and complicated intellectual terrain of defining what science is. As has been discussed in previous chapters, plaintiffs in both cases provided many witnesses to address what science is, and to show that evolution is science. The *McLean* plaintiffs' attempt to address what science is has been discussed in detail in this work. In *Kitzmiller,* plaintiffs' witnesses charged with addressing the definition of science gave similar answers. They stressed that scientific explanations must be testable, restricted to natural causes, tentative, and falsifiable. Particularly, as Scott discussed, they hit on the point that scientific explanations are restricted to natural causes, or methodological naturalism as it was called in *Kitzmller.*

This type of testimony, however, caused the *Kitzmiller* court to fall into the same intellectual and ethical dilemma that befell *McLean*. Although *Kitzmiller* engaged in a much less expansive and less-ambitious analysis of what science is than *McLean*, trying to confine its focus mainly on certain key attributes that qualify evolution as such, but disqualify intelligent design, the court purposely or negligently also simplified the intellectual problems associated with the demarcation of science from nonscience. The *Kitzmiller* court, as *McLean* before it, did not have the academic background, credentials, or proper forum to enter an intellectually complicated and specialized area such as demarcation, nor should it have done so.

Kitzmiller did, however, have the capacity to analyze the ID Policy's constitutional viability, using *Lemon* and the endorsement test, and to find it wanting. That should have been the end of the matter, rather than engaging in judicial overreach that brings into question the court's competence, integrity, and ultimate wisdom. I do not subscribe to Scott's position that demarcation of science from scientific creationism and ID had to be done. I do not believe the *Eppperson, McLean, Edwards,* and *Kitzmiller* courts had to do such an analysis when the primary religious purpose of creation science or ID can be clearly shown without engaging in a less-than-intellectually honest demarcation analysis. Applying the law of the appropriate time period, or a reasonable and judicially sound extension of that law, could have allowed the various courts to arrive at a sound, judicially constrained, and proper decision striking laws requiring balanced treatment for creation science or ID as violative of the Establishment Clause. After all, teaching creationism in a science class isn't' akin to supplying nonreligious textbooks to Parochial schools! It is distressing that the response of the courts to the thinly veiled attempts by creationists to present their religious views in a dishonest and misleading package was an equally dishonest effort to demarcate science from nonscience.

So we conclude our metaphorical journey through the history of the evolution in public-school cases with these comments. As we have previously discussed since *Scopes,* creationists have attempted to insert their particular brand of religion into the public-school science curriculum. These attempts coincided with evolutionary science moving away from a goal-directed, teleological view of nature. Although Darwin proposed, with natural selection, a materialistic, naturalistic mechanism for evolutionary change in *On the Origin of Species* in 1859, this view for the most part was not accepted at that time, and even Darwin was inconsistent in this matter. Except for a few evolutionists such as Thomas Huxley, evolution retained the progressive, teleological elements that had been associated with it since the Enlightenment, before Darwin's *Origin*. In the 1930s and 1940s things began to change. For example, George Gaylord Simpson during that time period consistently emphasized the more materialistic implications of evolution, stressing that there was no sign of purpose in nature. Simpson wrote that progress was an illusion because it was virtually impossible to define it meaningfully in biological terms. He believed that the human race was merely the end product of a series of essentially unpredictable natural events. Simpson rejected the concept that human moral values are reflected in any way in the laws of nature.

As previously discussed, Peter Bowler asserted, in the same vein, that although these elements are contained in the divergent, open-ended aspects of original Darwinian selection, they were generally rejected or ignored in the late nineteenth century and pre-World War I twentieth century. The emphasis on chance and unrepeatability (often termed contingency) in evolution has continued and developed during the latter part of the twentieth century and into the first decade of the twenty-first century. More recently, Stephen Jay Gould argueds that contingency represent evolution's deeper meaning and the controlling power in setting the pattern of life history. He suggested that almost every interesting event in life history falls outside of any boundary of predictability and falls into the realm of chance and unrepeatability, including the evolution of humans. An example that supports this view of nature, as we have discussed, is Alvarez's 1980 proposal that chance events, such as a large meteors or comets striking the earth, play determinative roles in shaping life history. Alvarez believed that his group documented at least one such impact, resulting in a mass extinction of most species on earth approximately sixty-five million years ago. Gould generally accepted Alvarez's ideas, stating that this extinction is of particular importance to humans because it included the elimination of dinosaurs and other related reptilian species. As a result, mammals were given an opportunity to radiate into the empty ecological niches, giving our own species an opportunity to ultimately evolve.

This view asserts that the history of life owes its shape, most importantly, to the differential success of groups surviving mass extinctions due to chance events. The implications of Alvarez's idea raise important questions concerning which species survive the catastrophe and which do not. For example, as we discussed, paleontologist David Raup questioned whether species survive most importantly because of genes (adaptive features) or chance. The event (catastrophe) that triggers the mass extinction obviously quickly and radically changes the conditions for survival in ways that are physiologically and anatomically unanticipated by organisms. Under the new post-catastrophe conditions, previously adaptive traits may no longer confer any adaptive advantage (promote survival), whereas traits with no previous adaptive significance may become most important for survival. This suggests that many species that survive the catastrophe do so only because of chance.

This reliance on contingency and a rejection of teleology makes it difficult to incorporate a supernatural being or divine plan into life history. Understanding this provides another explanation for the refusal of many scientists, teachers, lawyers, and judges to accommodate any type of creationism or supernatural intervention or guidance in the teaching of origin of the species or life history in public schools.

As previously stated, neither creation science, nor its most recent incarnation, intelligent design, is without deep religious underpinning and deep religious roots. Their religious nature is irrefutable and underscored by their own literature and the testimony of their proponents and experts in various court cases. This point is driven home by the incredibly weak defense mustered by expert witnesses for intelligent design in the *Kitzmiller* case and the incredibly weak defense mustered by expert witnesses for creation science in *McLean*.

Collectively and individually, their testimony casts doubt on intelligent design or creation science as an alternative explanation to evolution for life history and underscores their religious nature.

However, this doesn't adequately explain why mainstream religions, for the most part, were also against creationist's attempts to insert their brand of religion into the science curriculum with many, as we have seen, joining as plaintiffs in key cases. Perhaps the same type of competition is at work here that Ronald Numbers suggested as a reason mainstream science objects to creationism, "elimination of a politically powerful competitor for scarce resources." In this case, money and members. Economic explanations in capitalistic societies for political and even religious actions are not unreasonable.

I suppose the fact that there is continuing and increasing secularization of society at all levels and among all mainstream institutions, including the mainstream religions, also has relevance. Please think about the following for a minute as illustrative of this point. During the COVID-19 pandemic, the mainstream institutions turned their faces most profoundly toward science. Of course, I as a scientist don't object to relying on science and scientific research to achieve a socially beneficial result. A reliance on science as a way of preserving life and combating all types of disease is well-established and historically worthy of trust and belief. But I ask, where was, and is, the reliance on God or the supernatural equally pervasive and certain? It generally wasn't found in the mainstream synagogues, mosques, and churches during this crisis. I saw little outreach to people by the mainstream religious institutions. I saw little belief exhibited or expressed by them that a divine intervention is possible. Some politicians even made it explicit that our hope was only science. Society, in all of its parts, even its religious institutions, is highly secularized.

However, whatever the reasons for opposition to creationism, it must be admitted that balanced treatment laws mandating the inclusion of creationism into the public-school science curriculum carries problems beyond the obvious. One is the frontal attack on traditional values of academic freedom that it promotes. This was made explicit in the *Edwards* case. As the *Edwards* court stated, academic freedom is almost universally defined as a right by teachers to teach what they choose concerning their particular subject. Yet creationists urged the *Edwards* court to redefine it to require teachers to teach creation science if evolution is taught. This was done under the pretext of providing students with more diversity in their education. However, if that was creationists' real goal and purpose, wouldn't they want more than two renditions of how we got here be taught? After all, doesn't every culture and religion have their own story to tell?

Yet aren't the efforts of creationists to limit academic freedom part of a more general trend in our society from both religious and secular, left and right? We see, at almost every level of our society, efforts on the part of government and private institutions to limit freedom and free expression of thought. Oftentimes, labels are attached to laws and ideas that suggest the opposite of the real motive and intent: for example, the Patriot Act. There is little that's patriotic as to how it's been implemented and abused in so many instances by the government. Or more recently, the Inflation Reduction Act, which despite much good that it contains, will do little to reduce inflation based upon the economic analysis of most economists. We are increasingly living in an *Orwellian Nineteen Eighty-Four* reality.

However, doesn't the evolution in public-school cases also show that many in opposition to creationism also have little respect for freedom of thought or opposite views? For example, remember Gould's statement, "All thinking people accept the biological fact of evolution" and that "life on earth is not the result of special creation." Is he correct? As we have previously discussed, polls show that a large proportion of individuals in our society don't accept evolution; among them are at least some judges and scientists. Isn't Gould's criterion for being a "thinking" person somewhat disingenuous and limited?

Also recollect the 1994 Ninth District Federal Court of Appeals case, *Peloza v. Capistrano School District*, where an Appeals Court upheld a lower court decision allowing school districts to require science teachers to teach evolution in biology class, no matter what a teacher's feeling on the subject. Where is the academic freedom as defined by Edwards ("right by teachers to teach what they choose concerning their particular subject.") in this decision?

The erosion of traditional concepts of academic freedom has ramifications throughout the American educational landscape and makes public education even more vulnerable to politics than it is currently. If legislatures can determine what is taught in biology, then what will be next? Chemistry, physics, history, political science, geography, business, or maybe even math? The problem created is obvious and palpable. For example, as correctly articulated in *McLean,* evolution permeates public-school textbooks and teaching, throughout various intellectual disciplines, not just biology. Other subjects, such as world history, geology, zoology, botany, psychology, anthropology, sociology, philosophy, physics, and chemistry all contain evolutionary ideas. Mandating that creationism be taught in all of various intellectual topics would make public education religious: more precisely, evangelical or Fundamentalist religious. This brings us back to an important point. Application of the third prong of the *Lemon* test, mandating that a law cause no excessive entanglement between church and state, would have been sufficient to strike down creationist or ID statutes without courts engaging in demarcation analysis and other judicial and intellectual overreaches.

Therefore, none of the issues raised allow courts and the mainstream scientific establishment to be excused for their approach to the evolution cases since *Epperson.* As we have seen, the courts for the most part have consistently accepted that evolution is science without any question or analysis. This in itself understandable; courts and judges don't exist in a vacuum, and they must take judicial notice of the obvious, and it is obvious that evolution is science. Judges make common-sense rulings based upon their life experiences all the time. However, misrepresenting complex demarcation, philosophical, and scientific ideas and denigrating opposing ideas for ideological reasons—as we have seen demonstrated in certain cases—is not acceptable.

Three important examples of this are seen in *Epperson, McLean,* and *Kitzmiller.* In *Epperson,* we not only see a Supreme Court accepting mainstream science's view of evolution without analysis and asserting that creationism is merely an anachronism that is indefensible in light of modern scientific knowledge. How is it beneficial to denigrate the religious beliefs of a significant portion of the American population? Remember, the selective and biased use of the legislative history and history of the law by the *Epperson* court ignored that two-thirds of the Arkansas voters, far more than can be attributed solely to fundamentalism, supported that Law. In addition, it ignored that supporters of the law were not homogeneous in their thoughts, but they had mixed motives. For example, Bryan, in participating in the crusade that ultimately resulted in the enactment of both the Tennessee and Arkansas laws was equally concerned with preventing war and social exploitation as he was religion. Remember, Bryan was convinced that these social ills followed unfailingly from Darwinism. Further, as Ronald Numbers pointed out, studies have shown that the people who joined Bryan's antievolution crusade came from all walks of life and all parts of the country. In essence, the *Epperson* court had little documentation for its conclusions about the purpose of the Arkansas law and had no legitimate reason to denigrate the views of what may have been a majority of the Tennessee and Arkansas citizens. It did so for ideological reasons to further its particular anticreationist agenda.

Remember also the accusation that Justice Fortas altered the meaning of an advertisement the court used to support a sole purpose interpretation of the statute. The unaltered version supports a different stated purpose; that is, promoting religious neutrality. Bryan, a proponent of the act, did not believe creationism should be legally taught in public schools; therefore, he sought neutrality by also removing evolution.

Yet a more constrained and fair reading of the legislative and political history of the Arkansas Act would have allowed the *Epperson* court under the existing purpose and effect test of that time (*Abington School District v. Schempp*), or a reasonable extension of it, to invalidate the Arkansas law as predominantly religious in nature and effect without engaging in the duplicity and overreaching of the *Epperson* decision.

In *McLean,* we see a similar situation to *Epperson* where the court accepted evolution, *carte blanche,* as science yet rejects the scientific status of creation science based upon demarcation criteria that were disingenuous at best and deliberately misleading at worst. *McLean* also simplified, and perhaps outright misled, as to what subject matter is contained within evolutionary biology contending that origin-of-life questions are precluded. In doing so, the court entered into intellectual terrain that it had no competence to enter. As has been discussed, these questions are not quite as simple as the court would have us believe. *Kitzmiller,* in many ways, is the mirror image of *McLean,* except with intelligent design as the subject matter instead of creation science.

I reiterate an idea made more than once previously. Good evidence shows that creationists knew they were attempting to counterbalance evolution with ideas that were fundamentally religious in nature. Yet they persisted despite the legal setbacks, and they cynically changed their tactics to further their goal. There is no ethical, intellectual, political, or legal honesty to be found in their motives and actions. I share the belief of most federal courts that have heard the evolution in public-school cases that creationism or intelligent design does not belong in any public-school science curriculum. Yet as we have seen, the courts that adjudicated the evolution in public-school cases over the years frequently also did not have a commitment to procedurally fairness in their decision-making. I wonder again, does this issue serve as a model demonstrating a much-larger and more-pervasive problem with many of our institutions and the political and legal systems?

This brings me to the last question raised: is there any chance for accommodation? We live in a society that increasingly sees compromise as something politically or ethically objectionable and undesirable. This isn't a time in the history of this nation where compromise is considered positive. We are a polarized society, increasingly so, in almost every way. It's no surprise that this polarization and factionalism has permeated almost all aspects of life. I have witnessed this change with interest and mixed feelings in almost every part of the political and social arena.

To digress, I see it in many ways as positive; finally, structural injustices in our society seem to be being addressed partly because of this change in ways that might finally bring a true bend toward greater justice and equity. Yet I wonder, despite all of this, are we still ignoring and being conditioned to not address the most important structural inequity in our society? That is, severe economic disparity? Can there ever be justice in a society that has embedded in its very structure the economic disparity we find in ours?

This issue does not seem to be adequately addressed, as corporations espouse social justice and an end to racism, all good things to support, while they pay their CEOs hundreds of times the amount a factory worker or service worker makes. There is unequal access to higher education, unequal access to elite colleges and universities, unequal access to quality high schools and unequal distribution of government funds to schools. There is unequal access to medical care and legal protection, inequitable sentencing and penalties under the law. I could go on and on. Many of these inequities are because of structural racism, all of them are because of unequal and inequitable distribution of wealth. Which few corporations, which few politicians, are speaking out against this? Wealth in our society carries privileges that most everyone seems to accept. Being leaders of mainstream groups and institutions also seem to carry the same privileges. Perhaps that's why during the COVID-19 pandemic, politicians could rightfully demand masks and lockdowns for citizens, yet violate the rules when they found their rules personally inconvenient. Perhaps that's why wealthy businesspersons and politicians can rightfully demand austerity and restraint from citizen concerning environmental issues and climate change, yet own and live in multiple homes of gigantic proportions, own and drive multiple luxury vehicles, own yachts, and even private jets. We have become so acclimated to these incredible hypocritical behaviors that most of us see it as natural or consider it part of the normal structure of our society.

I was traditionally trained as an attorney to respect compromise, and I was taught that it was necessary for the reasonable, efficient, and even equitable functioning of the political and legal systems. However, I have increasingly questioned this philosophy. Structural inequities in the system are often propped up, bolstered, and supported because we are too willing to compromise ethical and equitable issues. Oftentimes, the compromises themselves are self-serving and reflect, ratify, and support the structural inequities of the system. However, there is a legitimate other side to this issue. Lack of compromise can easily lead to social chaos and political factions that destabilize society and political institutions in unproductive ways for everyone. I have come to an uncomfortable position that one must distinguish those issues where compromise will promote the common good, social progress, and justice from those issues where compromise is not ethically possible and merely contributes to maintaining prevailing structural inequities and status quo.

Distinguishing which issues falls into which category isn't always easy intellectual or ethical decision-making. One thing I noticed about this society is that there is a great deal of obfuscation about the core ethical, political, and economic issues underlying the social inequities and injustices. For example, as indicated above, often leaders of industry and politicians appear to support social-justice issues of one type or another yet ignore or actively oppose addressing core issues such as economic disparity and lack of economic justice, which serve as the structural foundation of societal injustice. As long as economic injustice prevails, there can never be just and equitable access to medical care, legal protection, educational opportunity, and other societal benefits and opportunities.

So we leave our digression and return to the problem at hand. Is there a place for creationism in the public-school science curriculum? Is this an issue that can be compromised? Can the opposing parties find an accommodation with one another? Will creationism's inclusion into the science curriculum in some manner promote greater justice? I feel sympathy for creationist parents of public-school children, or the students themselves, who rebel against a government-run system that directly challenges a core religious belief. Yet the solution can't be inclusion of a particular brand of Christian religion, or for that matter any brand of Christian religion, into the public-school science curriculum. Our society is far too diverse now and modern constitutional interpretation far too sensitive to First Amendment religious issues to allow that. At this point there are so many factors that argue against inclusion of creationism. I see no place for it in the public-school science curriculum.

There are several reasons why I write this. First, inclusion would cause a significant financial expenditure on the part of the government at the local and federal levels and require constant and in-depth oversight over almost the entire public-school curriculum. Evolution is not just another topic in biology. It is biology's main unifying concept. It's the intellectual glue that holds modern biology together into a coherent discipline, so it runs through all areas of biology. In addition, as we have seen, evolution permeates and percolates through so many other intellectual pursuits—chemistry, physics, philosophy, economics, and anthropology—just to name a few. The economic cost of equal time or balanced treatment for creationism would be high. The time expenditure to do so would be even more prohibitive. The necessary revisions of textbooks alone would be enormous. The bureaucracy needed to monitor the textbooks, curricula, and teaching for balanced treatment implementation would be untenable. In essence, local, state, and federal intrusion into the educational process of public schools would be unlike anything ever attempted or accomplished in the history of American education. It would result in actions and behavior that would be the very definition of "excessive government entanglement" with religion. And perhaps, most importantly, it would be a frontal attack on the academic freedom of teachers and instructors of the various disciplines affected.

Second, creationism and intelligent design run against an increasingly secular trend in the United States. We are now a very secular society. In addition, we are becoming an increasingly diverse society; Christianity does not hold the hegemony that it once did in the United States even a short time ago. How do you entertain Christian dogma, let alone that from a splinter group (albeit a large splinter group) within Christianity, in the science classroom in modern-day America with its increasing diversity? Why not teach the Koran's story of origins, or one of the Hindu stories, or one of the Native American origin stories, or even theistic evolution? In twenty-first-century United States, any one or all of these groups, or a host of other ones, have as much right as do Christian creationists to demand that their origin stories be taught. It's not constitutionally feasible, or feasible in any way, to make this type of accommodations.

Third, evolutionary biology, as we have discussed, has moved a considerable distance since *Scopes* from its progressive, teleological, Enlightenment roots and it has generally embraced contingency, rejected progressivism, and rejected moral principles in nature. Evolution, at this time, is not a science able to share space with teleological, morality-based philosophy or religion such as creationism.

Fourth, there is an increasing pervasive tendency in our society to see the elimination of opposing political, ideological, and religious views as positive. The traditional understanding, "I don't agree with your position or what you are saying, but I will defend your right to have it or say it," increasingly no longer seems to apply in the United States. I see this as an incredibly negative trend. I think that the story told in this work makes it clear that a philosophy of accommodation and tolerance did not, and does not, exist with either side of the cases discussed. The almost hundred years of animosity between the two sides is palpable. Unfortunately, this lack of willingness to compromise seems to reflect a more general trend in our society found in art, music, politics, etc.

Finally, as we have discussed, there are genuine First Amendment Constitutional issues that can't be ignored by inclusion of creationism into the science curriculum. Not only "entanglement" issues, but also purpose and effect issues that cause the creationist position to fail a *Lemon* or endorsement test as they were understood and interpreted.

Where does this leave us? A recent survey shows an increasing ascendency of evolution in the United States public-school science curriculum since *Kitzmiller*. It appears that a multifaceted attack on inclusion of creationism in the public schools is having an effect on the curriculum. Eric Plutzer,[1] Glenn Branch, and Ann Reid provided evidence for this by comparing a 2007 survey with a comparable one done in 2019 (see https://evolution-outreach.biomedcentral.com/track/pdf/10.1186/s12052-020-00126-8.pdf).

In 2007, less than two years after *Kitzmiller*, the first national survey concerning the teaching of evolution showed that only about one-third of the public high school biology teachers presented evolution consistent with the recommendations of United States scientific establishment. About 13 percent of public-school teachers taught that creationism is a valid scientific alternative to modern evolutionary biology. However, things changed radically within twelve years after that survey. The 2019 study showed that in the twelve years between the two surveys, evolution instruction in US public high schools changed substantially. Far fewer teachers report sending mixed messages, and many more report emphasizing to their students "the broad consensus that evolution is a fact, even as scientists disagree about the specific mechanisms through which evolution occurred." In addition, teachers report devoting substantially more class hours to evolution, including human evolution. New teachers who entered the profession after 2007 are especially diligent in this regard, but both senior and new teachers are showing a change. Clearly, creationism in public-school science classes is significantly in decline.

The reasons for this change are numerous. Not only did *Kitzmiller* demonstrate that creationism—in whatever form it took, including intelligent design—faced insurmountable barriers to achieve a place in public school science education, but organizations including the National Science Teaching Association, the National Association of Biology Teachers, and the National Academy of Sciences produced statements, reports, classroom resources, and professional development opportunities to advance the effective teaching of evolution in the public schools. In addition, the development in the 2010s of the Next Generation Science Standards (NGSS) afforded a central place to evolution in biology education. By 2019, these standards were adopted in twenty states plus the District of Columbia. Twenty-four additional states have developed standards on the same framework (National Research Council 2012) on which the NGSS is based and therefore recommend a comparable treatment of evolution in the public-school science curriculum. The NGSS Standards have been increasingly reflected in textbooks, online resources, preservice teacher education, and in-service teacher professional development opportunities, with the National Science Teaching Association firmly endorsing them (https://evolution-outreach.biomedcentral.com/track/pdf/10.1186/s12052-020-00126-8.pdf).

Nonetheless, although these initiatives are important in moving United States public schools toward greater intellectual integrity in teaching biology in public schools, they do not address in any significant manner the fears and concerns of creationists who feel as though there is a direct attack on their religious beliefs and their ability to pass them down to their children.

So where do we go from here? Time will tell. Perhaps a partial solution that might lead to some accommodation and a decrease in animosity would come from willingness on the part of scientists and science teachers to emphasize and embrace an attribute that they often claim is part of modern science—its tentative nature. Often, evolution is taught as fact, written about as fact, spoken about as fact. Yet despite this, most scientists and science teachers claim that science is tentative. We are taught that science isn't certain. We are told that religion is certain, science is tentative.

I don't claim that teaching evolution as fact makes it religion, but I do submit that it makes it more ideological. If creationism fails as science, the teaching of evolution as fact fails as a sound and ethical teaching methodology. Yet the courts since *Epperson* have accepted and embraced this ideology without question.

I am not suggesting that the courts interfere with academic freedom by indicating how evolution is taught; I am suggesting that science teachers and scientists better understand the ideological implications of teaching evolution as fact and the ethical consequences of doing so. This might be a point where accommodation, compromise, and the United States Constitution come together.

As this book was being edited, a Supreme Court case was decided that may have potential impact on future evolution in public school cases in unpredictable ways. The case is a First Amendment case, a "religious activities within public schools" case to be specific. Here are the relevant facts. Joseph Kennedy lost his job as a high school football coach because he knelt at midfield after games to offer quiet prayers of thanks to God. He also did various and other sundry religious practices at other times. As a result, the school district disciplined him, and he ultimately left his position. The basis for the school district's discipline was their belief that Kennedy's action violated the First Amendment Religious Clauses because the school district believed his actions endorsed Kennedy's religious beliefs. This matter was appealed by Kennedy to the appropriate district court, which, using endorsement and *Lemon* test analyses, found in favor of the school district. The district court's decision was appealed by Kennedy to the Ninth Circuit Federal Appeals Court, which affirmed the district court's decision. Ultimately, Kennedy appealed the case to the United States Supreme Court, which agreed to hear it in *Kennedy v. Bremerton School District* (see October Term, 2021, https://www.supremecourt.gov/opinions/21pdf/21-418_i425.pdf)

On June 27, 2022, the Supreme Court issued a written decision holding that the lower courts and the school district's reasoning was "misguided." The Supreme Court concluded that the Free Exercise and Free Speech Clauses of the First Amendment protect expressions like Kennedy's and that the Establishment Clause does not require the government to single out "private religious speech for special disfavor." The court held that the Constitution and the best of our traditions—"counsel mutual respect and tolerance, not censorship and suppression, for religious and nonreligious views alike."

The court then unambiguously overruled the *Lemon* and the endorsement tests, holding that in place of *Lemon* and the endorsement test, the Establishment Clause must be interpreted by "'reference to historical practices and understandings.'" The court stated that an analysis focused on original meaning, and history is a proper one.

Although the subject matter of *Bremerton School District* is not on point with the First Amendment evolution in public school cases analyzed in this work, the case still has importance here. One reason is that as we have seen throughout this work, since *Epperson*, every evolution case has been analyzed on First Amendment grounds with the majority of courts involved using the *Lemon* test and with the *Kitzmiller* court using both the *Lemon* and the endorsement tests to evaluate the constitutionality of the laws or ruling, seeking to limit or prohibit the teaching of evolution in public schools.

It will be interesting to see how, or if, *Bremerton Scholl District* will affect future interpretations of how the First Amendment will be applied to evolution in public school cases perhaps to come.

At this point, we are at the end of this work. It's been an interesting intellectual journey. Somehow, I have a feeling that this story is far from finished.

ENDNOTES

Introduction

1 https://plato.stanford.edu/entries/galileo

2 http://law2.umkc.edu/faculty/projects/ftrials/galileo/galileoaccount.html

3 https://newsroom.ucla.edu/releases/the-truth-about-galileo-and-his-conflict-with-the-catholic-church

4 https://www.britannica.com/biography/Galileo-Galilei

5 *Scopes v. State*, 154 Tenn. 105, 289 S. W. 363 (1927).

6 https://www.gutenberg.org/files/1228/1228-h/1228-h.htm

7 Frederick Sproull, *Reflections on Evolution*, (Page Publishing, New York, NY, 2017).

8 Theodosius Dobzhansky, "Biology, Molecular and Organismic," *American Zoologist* 4, (1964).

9 Richard C. Lewontin, *It Ain't Necessarily So*, (London: Granta Books, 2000).

10 Stephen J. Gould, "Spin Doctoring Darwin," *Natural History*, 7/1995.

11 Steve Mirsky, "Influence of Darwin on Modern Thought," *Science American*, November 24, July 2009.

12 Ernst Mayr, *The Growth of Biological Thought*, (Cambridge, Mass.: Belknap P. of Harvard University Press, 1982).

13 Ernst Mayr, *One Long Argument*, (Cambridge, Mass.: Harvard University Press, 1991).

14 Ernst Mayr, *Toward a New Philosophy of Biology*, (Cambridge Mass.: Harvard University Press, 1988).

15 Stephen J. Gould, *Hen's Teeth and Horse's Toes*, (New York: W. W. Norton, 1983).

16 https://www.huffingtonpost.com/2014/06/02/creationism-america-survey_n_5434107.html

17 https://www.huffingtonpost.com/2014/06/02/creationism-america-survey_n_1571127.html

18 Larry A.Witham, *Where Darwin Meets the Bible*, (New York: Oxford Press, 2002).

19 http://www.msnbc.msn.com/id/19956961/ns/world_news-europe/t/pope-creation-vs-evolution-clash-absurdity

20 http://www.newsweek.com/pope-franciss-remarks-evolution-are-not-controversial-among-roman-catholics-281115.

21 http://arn.org/docs/johnson/pjdogma1.htm

22 Douglas J. Futuyma, *Science on Trial*, (New York: Pantheon Books, 1982).

23 https://www.science.org/doi/epdf/10.1126/science.274.5288.717

24 Ronald Numbers, *The Creationists*, (New York: Knopf, 1992).

Chapter 1

1 *Scopes v. State*, 154 Tenn. 105, 289 S. W. 363 (1927).

2 https://www.gutenberg.org/files/1228/1228-h/1228-h.htm

3 Ronald Numbers, *The Creationists*, (New York, Knopf, 1992).

4 Edward J. Larson, *Evolution: The Remarkable History of a Scientific Theory*, (New York: Random House, 2004).

5 Peter J. Bowler, *Evolution: The History of an Idea*, (Berkley: University of California Press, (1989).

6 *Epperson v. Arkansas*, 393 U.S. 97, (1968).

7 Judith A. Villarreal, "God and Darwin in the Classroom: The Creation/Evolution Controversy," 64 *Chi-Kent L Rev.*, 335 (1988).

8 Stephen J. Gould, *Hen's Teeth and Horse's Toes*, (New York: W. W. Norton, 1983).

9 https://www.jstor.org/stable/1073701

10 https://www.aclu.org/other/aclu-history-scopes-monkey-trial

11 Edward J. Larson, *Summer for the Gods*, (New York: Basic Books 1997).

12 Stephen J. Gould, "William Jennings Bryan's Last Campaign," 77 *Nebraska History* 177 (1996).

13 Stephen J. Gould, *Flamingo's Smile*, (New York: Norton 1985).

14 Stephen J. Gould, *Mismeasure of Man*, (New York: Norton 1980).

15 *Buck v. Bell*, 274 U.S. 200 (1927).

16 http://faculty.uca.edu/benw/biol4415/papers/carriebuck.pdf

17 Peter J. Bowler, *Eclipse of Darwinism*, (Baltimore: Johns Hopkins University Press 1983).

18 Peter J. Bowler, *Darwinism*, (Woodbridge: Twayne Pub.1983).

19 Tenn. Code Ann. Section 27-1-113(1925).

20 https://www.aclu.org/other/state-tennessee-v-scopes

21 *Cantwell, et al. v. Connecticut*, 310 U.S. 296 (1940).

22 *Everson v. Board of Education*, 330 U.S. 1 (1947).

23 Stephen J. Gould, Hen's *Teeth and Horse's Toes*, (New York: W. W. Norton, 1983).

Chapter 2

1   Stephen J. Gould, *Hen's Teeth and Horse's Toes*, (New York: W. W. Norton, 1983).

2   Judith A. Villarreal, "God and Darwin in the Classroom: The Creation/Evolution Controversy," 64 *Chi-Kent L Rev.*, 335 (1988).

3   *Epperson v. Arkansas*, 393 U.S. 97 (1968).

4   Ark. Stat. Ann. Sections 80-1627-1628 (1960 Repl. Vol.).

5   Tenn. Code Ann. Section 27-1-113(1925).

6   *State v. Epperson*, 416 S.W. 322 (1967).

7   *Scopes v. State*, 154 Tenn. 105, 289 S. W. 363 (1927).

8   https://uscivilliberties.org/themes/3136-application-of-first-amendment-to-states.html; see also *Cantwell v. Connecticut*, 309 U.S. 626 (1940) and *Everson v. Board of Education*, 330 U.S. 1 (1947).

9   *Abington School District v. Schempp*, 374 U.S. 203 (1963).

10 L. B. Halpern, "Edwards v. Aguillard: "The Supreme Court Evaluates the Sincerity of the Louisiana Legislature," 34 *Loy. L. Rev.* 406 (1988–1989).

11 George E. Webb, *The Evolution Controversy in America*, (Lexington: University Press of Kentucky, 1994).

12 Ernst Mayr, *The Growth of Biological Thought*, (Cambridge, Mass.: Belknap P. of Harvard University Press, 1982).

13 Ernst Mayr, *One Long Argument*, (Cambridge (Mass.: Harvard University Press, 1991).

14 Edward O. Dodson, Evolution: Process and Product: (Boston: Prindle Weber & Schmidt, 1976).

15 Peter J. Bowler, *Evolution: The History of an Idea*, (Oakland: University of California Press, 1984).

16 Mark Ridley, *Evolution*, (Hoboken: John Wiley & Sons, 1996).

17 Peter J. Bowler, *Eclipse of Darwinism*, (Baltimore: Johns Hopkins University Press 1983).

18   Ernst Mayr, *Toward a New Philosophy of Biology*, Cambridge (Mass.: Harvard University Press, 1988).

19   Douglas J. Futuyma, *Evolution*, Sunderland, MA: Sinauer Associates (4th), 2017.

20   Ronald Numbers, *The Creationists*, (New York: Knopf, 1992).

21   Edward J. Larson, *Evolution: The Remarkable History of a Scientific Theory*, (New York: Random House, 2004).

22   Edward J. Larson, *Summer for the Gods*, (New York: Basic Books 1997).

23   *Edwards v. Aguillard*, 482 U.S. 578 (1987).

Chapter 3

1   Peter J. Bowler, Evolution: The History of an Idea, (Berkley: University of California Press, 1989).

2   Lucien J. Dhooge, "From Scopes to Edwards: The Sixty-Year Evolution of Biblical Creationism in the Public School Classroom," 22 *University of Richmond Law Review* Volume, Issue 2, Article 6 1988).

3   T.M Serra, "Louisiana's Balanced Treatment Act is Facially Invalid because Its Primary Purpose Is to Promote a Particular Religious," 65 U. Det. L. Rev. 843 (1987–1988).

4   S. Shaeffer. "Edwards v. Aguillard: Creation Science and Evolution—the Fall of Balanced Treatment Acts in the Public Schools," 25 *San Diego L. Rev.* 829 (1988).

5   *Wright v. Houston Independent School District*, 366 F. Supp. 1208 (1972) and 486 F.2d 137 (1973).

6   *Daniel v. Waters*, 515 F2d. 485 (1975); see also https://ncse.ngo/daniel-v-waters-and-steele-v-waters-1973-1975

7   Tenn. Code Ann. Section 49-2008(1)-(2) (1973).

8   *Lemon v. Kurtzman*, 403 U.S. 602 (1971).

9   *Abington School District v. Schempp*, 374 U.S. 203 (1963).

Chapter 4

1   *McLean v. Arkansas*, 529 F. Supp. 1255 (1982).

2   Ark. Stat. Ann. Sections 80-1663 to 1670 (Supp. 1981).

3   *Lemon v. Kurtzman*, 403 U.S. 602 (1971).

4   Frederick Sproull, *Reflections on Evolution*, (Page Publishing, New York, NY, 2017).

5   Douglas J. Futuyma, *Evolution*, Sunderland, MA: Sinauer Associates, (4th), 2017

6   Stephen J. Gould, *Bully for the Brontosauras*, (New York: W. W. Norton, 1991).

7   Lucien J. Dhooge, "From Scopes to Edwards: The Sixty-Year Evolution of Biblical Creationism in the Public School Classroom," 22 *University of Richmond Law Review* Volume, Issue 2, Article 6 1988).

8   *Stone v. Graham*, 449 U.S. 39

9   *Epperson v. Arkansas*, 393 U.S. 97 (1968).

10 https://www.researchgate.net/publication/285372385_The_scientific_status_of_modern_evolutionary_theory

11 Peter J. Bowler, *Evolution: The History of an Idea*, (Berkley: University of California Press, 1989).

12 Elliott Sober, *The Nature of Selection*, Chicago: University of Chicago Press, 1984.

13 Elliott Sober, *Philosophy of Biology*, Denver: Westview Press, 1993.

14 Stephen J. Gould, *Ever Since Darwin*, (New York: W. W. Norton, 1978).

15 I. Lakatos, "The History of Science and its Rational Reconstructions", in R.C. Buck and R.S. Cohen (eds.), PSA 1970: Boston Studies in the Philosophy of Science, 8, Dordrecht: Reidel, pp. 91–135.

16 K. R. Popper, "Natural Selection and the Emergence of the Mind," 32 *Dialectica*, 339-355, 1978

17 Ronald Numbers, *The Creationists*, (New York: Knopf, 1992).

18 Larry Laudan, "Science at the Bar," 7 *Science, Technology, and Human Values*, No. 41 (Autumn 1982) pages 16–19.

19 http://faculty.washington.edu/lynnhank/Ruse.pdf

20 http://www.antievolution.org/projects/mclean/new_site/pf_trans/mva_tt_p_ruse.html

21 *Edwards v. Aguillard*, 482 U.S. 578 (1987).

22 *Kitzmiller v. Dover Area School District*, 400 F. Supp. 2d 707 (2005).

23 http://www.antievolution.org/projects/mclean/new_site/pf_trans/mva_tt_p_ruse.html

24 https://www.lri.fr/~mbl/Stanford/CS477/papers/Kuhn-SSR-2ndEd.pdf

25 Richard Levins and Richard Lewontin, *The Dialectical Biologist*, (Boston: Harvard University Press, 1985) see also Richard Lewontin, *Biology as Ideology*, Boston: Harvard University Press, 1991).

26 Edward O. Dodson, *Evolution: Process and Product*: (Boston: Prindle Weber & Schmidt, 1976).

27 John Maynard Smith, *The Theory of Evolution*, (Cambridge: Cambridge University Press, 1993).

28 Lisa A. Urry, Michael L. Cain, Steven A. Wasserman, Peter V. Minorsky, Rebecca Orr, *Campbell Biology* (12th Edition), (London: Pearson, 2020).

29 Stephen J. Gould, *Wonderful Life*, (New York: W. W. Norton, 1989).

Chapter 5

1 *Edwards U.S. v. Aguillard*, 482 578 (1987).

2 La. Rev. Stat. Ann. Sections 17.286.1-17.286.7 (West 1982).

3 Lucien J. Dhooge, "From Scopes to Edwards: The Sixty-Year Evolution of Biblical Creationism in the Public School Classroom," 22 *University of Richmond Law Review* Volume, Issue 2, Article 6 1988).

4 *Aguillard v. Edwards*, 765 F. 2d 1251 (1985).

5 *Aguillard v. Edwards*, 778 F. 2d 225 (1985).

6 Mary E. Garvey, "Edwards v. Aguillard: The Supreme Court Evaluates the Sincerity of the Louisiana Legislature," 34 *Loy. L. Rev.* 406 (1988-1989).

7 *Lynch v. Donnelly*, 465 U.S. 668 (1984).

8 *Kitzmiller v. Dover Area School District*, 400 F. Supp. 2d 707.

9 *Wallace v. Jaffee*, 472 U.S. 38 (1984).

10 *Stone v. Graham*, 449 U.S. 39 (1980).

11 *Epperson v. Arkansas*, 393 U.S. 97 (1968).

12 S Schaeffer, "Edwards v. Aguillard: Creation Science and Evolution-the Fall of Balanced Treatment Acts in the Public Schools," 25 *San Diego L. Rev.*, 829, 1988

13 T.M. Serra, "Establishment Clause—Louisiana's Balanced Treatment Act is Facially Invalid because Its Primary Purpose Is to Promote a Particular Religious Belief," 65 *U. Det. L. Rev.* 843 (1987-1988).

14 M.A. Sares, "Edwards v. Aguillard: The Case of the Misplaced Purpose," 19 *U. Tol. L. Rev.* 727 (1987–1988).

15 *Edwards v. Aguillard,* Amicus Brief, National Academy of Sciences (umn. edu).

16 https://casetext.com/case/edwards-v-aguillard

17 https://digitalcommons.law.uga.edu/cgi/viewcontent.cgi?article=1502&-context=fac_artchop

18 https://digitalcommons.law.uga.edu/cgi/viewcontent. cgi?referer=&httpsredir=1&article=1502&context=fac_artchop

19 Michael Brant Shermer, "Science Defended, Science Defined: The Louisiana Creationism Case," 16 *Science, Technology, & Human Values,* 517 (1991).

20 Briefs filed at *Edwards v. Aguillard* (U.S. 1985) (No. 85-1513).

21 Brief of amici curiae, filed for the Catholic League for Religious and Civil Rights, et al., by Stephen Frederick McDowell, *Edwards v. Aguillard* (U.S. 1985) (No. 85-1513)

22 http://www.gallup.com/poll/21814/Evolution-Creationism-Intelligent-Design. aspx; http://www.cbsnews.com/news/poll-creationism-trumps-evolution/

23 http://www.huffingtonpost.com/2012/06/01/gallup-poll-americans-creationism evolution_n_1563800.html.

24 Stephen J. Gould, *Hen's Teeth and Horse's Toes,* (New York: W. W. Norton & Company, 1994).

Chapter 6

1 *Edwards U.S. v. Aguillard,* 482 578 (1987).

2 *Kitzmiller v. Dover Area School District,* 400 F. Supp. 2d 707.

3 Frederick Sproull, *Reflections on Evolution,* (Page Publishing, New York, NY, 2017); see also, Douglas J. Futuyma,) *Evolution,* Sunderland, MA: Sinauer Associates, (4th), 2017

4 *McLean v. Arkansas,* 529 F. Supp. 1255 (1981)., see also, https://www.science.org/doi/10.1126/science.215.4535.934

5 *Lemon v. Kurtzman,* 403 U.S. 602 (1971)

6 *County of Allegheny v. ACLU,* 492 U.S. 573 (1989).

7 James M. Lewis and Michael L. Vlid, "Controversial Twist of Lemon: The Endorsement Test as the New Establishment Clause Standard," 65 *Notre Dame Law Review,* Issue 4 Article 4 6-1-1999

8 Jesse H. Choper, "The Endorsement Test: Its Status and Desirability 18 *J.L. & Pol.* 499 (2002).

9   *Epperson v. Arkansas*, 393 U.S. 97 (1968).

10 *Daniel v. Water*, 515 F.2d 485 (1975).

11 Percival Davis and Dean H. Kenyon, *Of Pandas and People: The Central Question of Biological Origins*, 2nd Edition, (Boston: Haughton Pub Co; 2nd edition, 1993).

12 Barbara Forresor and Paul R. Gross, *Creationism's Trojan Horse: The Wedge of Intelligent Design* (1st Edition, Kindle Edition, Oxford: Oxford University Press, 2007).

13 Philip E. Johnson, *Darwin on Trial*, (Washington, DC, Regnery Gateway, 1991).

14 *Texas Monthly, Inc. v. Bullock*, 489 U.S. 1 (1989).

15 Stephen J. Gould, *The Panda's Thumb: More Reflections in Natural History*, (New York: W. W. Norton, 1980).

16 https://www.pewforum.org/2009/02/04/the-social-and-legal-dimensions-of-the-evolution-debate-in-the-us/

17 https://scholars.unh.edu/unh_lr/vol4/iss2/6/

18 *Stone v. Graham*, 449 U.S. 39, 1980

19 file:///E:/Book2%20and%20Book3%20-%2010-7-20/Articles/403-2147-1-PB.pdf

20 Michael J. Behe, *Darwin's Black Box: The Biochemical Challenge to Evolution*, (New York: Free Press, 1996).

21 Michael J. Behe, "Reply to my critics: A response to reviews of Darwin's Black Box: The Biochemical Challenge to Evolution," (16 (5) *Biology & Philosophy*, 685–709, 2001).

---------- ABOUT THE AUTHOR ----------

Frederick Sproull, PhD, JD, is a biologist and attorney. He received his PhD in developmental biology from the University of Pittsburgh. After which, he did postdoctoral research at Albert Einstein College of Medicine and the University of Pittsburgh School of Medicine. Currently, Dr. Sproull is chairperson of the Department of Biology at LaRoche University in Pittsburgh, Pennsylvania. Among the courses he teaches are evolution, population genetics, and an exploration of the Galapagos Islands. He has published a book titled *Reflections on Evolution*. Dr. Sproull is past president of the Pennsylvania Conference of the American Association of University Professors and past chairperson of the conference's committee. Dr. Sproull has a special interest in educational law.